T0340189

BASICS OF ENGINEERING TURBULENCE

BASICS OF ENGINEERING TURBULENCE

DAVID S-K. TING

Turbulence & Energy Laboratory
Centre for Engineering Innovation
University of Windsor
Windsor, Ontario, Canada

Amsterdam • Boston • Heidelberg • London
New York • Oxford • Paris • San Diego
San Francisco • Singapore • Sydney • Tokyo
Academic Press is an imprint of Elsevier

Academic Press is an imprint of Elsevier
125 London Wall, London EC2Y 5AS, UK
525 B Street, Suite 1800, San Diego, CA 92101-4495, USA
50 Hampshire Street, 5th Floor, Cambridge, MA 02139, USA
The Boulevard, Langford Lane, Kidlington, Oxford OX5 1GB, UK

Notices
Knowledge and best practice in this field are constantly changing. As new research and experience broaden our understanding, changes in research methods, professional practices, or medical treatment may become necessary.

Practitioners and researchers must always rely on their own experience and knowledge in evaluating and using any information, methods, compounds, or experiments described herein. In using such information or methods they should be mindful of their own safety and the safety of others, including parties for whom they have a professional responsibility.

To the fullest extent of the law, neither the Publisher nor the authors, contributors, or editors, assume any liability for any injury and/or damage to persons or property as a matter of products liability, negligence or otherwise, or from any use or operation of any methods, products, instructions, or ideas contained in the material herein.

Library of Congress Cataloging-in-Publication Data
A catalog record for this book is available from the Library of Congress

British Library Cataloguing-in-Publication Data
A catalogue record for this book is available from the British Library

ISBN: 978-0-12-803970-0

For information on all Academic Press publications
visit our website at http://store.elsevier.com/

Working together
to grow libraries in
developing countries

www.elsevier.com • www.bookaid.org

DEDICATION

The meaning of an endeavor is found in the process, more so than the final outcome. This book is dedicated to those who attempt to make the best out of everyday turbulence.

'Yesterday is history, tomorrow is a mystery, today is a gift of God, which is why we call it the present.' – Bil Keane

CONTENTS

LIST OF FIGURES

LIST OF TABLES

ACKNOWLEDGMENTS

In the absent of fore- and corunners, this book would have been but an anxious dream of starting a marathon without the stamina to cross the finish line. The author is particularly grateful to the strength from above and many individuals who eased this challenging endeavor, giving him the fuel needed to make it through the finish line. These instrumental individuals include the following:

Prof Dr D.J. Wilson, by whom the author was culturally shocked by flow turbulence in 1989. Some parts of this book have been written based on his lecture notes ("Mec E 632 Turbulent Fluid Dynamics," University of Alberta, 1989).

The numerous engineering artists who have supplied the beautiful figures. While their helping hands are explicitly recognized in the figure caption, a general heartfelt thank you goes to the Turbulence and Energy (T&E) Laboratory. Everyone who has contributed, one way or another, is a T&E-er at heart, even though some graduated before the official establishment of the T&E Lab. Also, thanks to many of my T&E colleagues – even Dr Rupp Carriveau, who stole the spotlight with his eloquent exaggerations in the foreword.

The Elsevier publishing team who is responsible for many sleepless nights. It started with Chelsea Johnston and Joseph P. Hayton. The nightmare became particularly real with Carrie L. Bolger. The extremely supportive reviewers are also responsible. In particular, Prof Dr Pierre Sullivan, Prof Dr Himanshu Tyagi, and the overly positive Prof Dr Alain deChamplain and Alain Fossi.

The Turbulent flow graduate classes of 2012, 2014, and 2015, especially those who picked up the many less-than-obvious errors scattered over the two versions of self-published lecture notes. During the final round, Jamie C. Smith went through each sentence and objected to countless "leap-of-faith" wordings, Sai Praneeth Mupparapu checked every tilde in the equations, and Hao Wu combed out whatever debris he came across.

Mom, dad, sisters, and brother, and Uncle Mitchell and the other founding members of the Allinterest Research Institute; fluid dynamics is still turbulenting my heart after all these years.

Naomi, Yoniana, Tachelle, and Zarek Ting for their unfailing love and continual encouragements, especially those conveyed via many philosophical sarcasms. They were there from the groundbreaking days of the first (self-published) version of this book with Naomi Ting's Books. It has been a long but rewarding journey together.

FOREWORD

Turbulence can be a very beautiful thing, a dynamic cascade of scales that connects us to each other and our environment. Occasions to study turbulence are abundant; examples flow from our bodies to the heavens. Opportunities for true understanding are considerably less ample. Sir Horace Lamb himself once famously quipped that on reaching heaven he hoped for divine enlightenment on just two matters: quantum electrodynamics and turbulence. He said he was "rather optimistic of the former." One barrier that has challenged more widespread understanding has been the lack of a true bridge to the topic. From the seminal works to most contemporary texts, the treatment of the subject is detailed and advanced. This is perfectly appropriate for select scholars of the science and sufficiently discouraging for the beginner, enthusiast, or cross-disciplinarian looking for application-level understanding. Subsequently, the ranks of the well-informed remain somewhat exclusive.

Enter David Ting. I have had the pleasure and challenge of working with David for the last 11 years. While David's turbulence publication record is impressive in its own right, I have always been more impressed by his dedication and concern for students. More than anyone I have ever known, he is able to simplify, rearrange, and relate complex matters to those lost sheep keen to join the flock of the initiated. When his conventional teaching toolset is not reaching the students, it is his unparalleled faith in them that inspires their personal development. I trust you will enjoy the bridge David has built with this textbook. If you will not subscribe to my endorsement, then please have faith; in the pursuit of turbulence enlightenment, it would seem a minimum requirement.

Rupp Carriveau

Dr Rupp Carriveau is the Associate Professor at the University of Windsor, Lumley Centre for Engineering Innovation. He is the coordinator in the Centre for Energy and Water Advancement and Director in the Turbulence and Energy Laboratory. Dr Carriveau serves on the Editorial Boards of Wind Engineering, Advances in Energy Research, and the International Journal of Sustainable Energy. He is the current President of the Underwater Energy Storage Society. He was recently designated as the University Scholar and has served as the Research Ambassador for the Council of Ontario Universities.

PREFACE

This book is intended for keen minds interested in flowing fluids. Specifically, it aims at removing the "fear of water" from those who are new to flow turbulence. The basic background on everyday flow turbulences, especially those encountered in engineering applications, forms the crux of the book. Some undergraduate knowledge of fluid mechanics and statistics is needed to best appreciate the material covered.

David S-K. Ting
August 14, 2015

PART 1

Some Basics of Flow Turbulence

CHAPTER 1

Introducing Flow Turbulence

The greatest achievements were at first and for a time dreams. The oak sleeps in the acorn.

– James Allen

Contents

Chapter Objectives

- To introduce the concept of flow turbulence.
- To learn about the fundamental characteristics of turbulent flows.
- To create a high-level appreciation of the pioneering explorations of turbulent flow.
- To preview the organization of the book.

NOMENCLATURE

g Gravity
h Height, (convective) heat transfer coefficient

Basics of Engineering Turbulence
http://dx.doi.org/10.1016/B978-0-12-803970-0.00001-5

L Characteristic length
m Mass
r Radius
Ra Rayleigh number, Ra = gravity/thermal diffusivity = $g\beta\Delta Th^3/(v\alpha)$
Re Reynolds number, Re = inertia force/viscous force = UL/v
s Spacing, gap
T Temperature, time period
Ta Taylor number, $Ta = \Omega^2 s^3 r_{inner}/v^2$
u Fluctuating velocity
U Velocity
$\mathbf{Bold}$ $\Rightarrow$ Tensor

Greek symbols

α Thermal diffusivity, thermal expansion coefficient
β Expansion coefficient
λ Wavelength
μ Dynamic (absolute) viscosity
v Kinematic viscosity
ρ Density
Ω Vorticity, $\Omega \equiv \nabla \times \mathbf{U}$

1.1 INTRODUCTION

In this introductory chapter, the importance of flow turbulence is established. To facilitate a conceptual understanding of turbulent flows, the indispensable fundamental characteristics are conveyed. This is followed by a brief historic account of the classical turbulence explorations to further the appreciation thereof. The chapter concludes with a brief outline of this precursory text on the marvel of flow turbulence.

Turbulent motion is the natural state of most fluids, whereas laminar flow is the exception and not the rule, in both nature and technology. Turbulence is an underlying mechanism in cloud formation and atmospheric transport (Weil et al., 1992; Vaillancourt and Yau, 2000; Grabowski and Wang, 2013), practical droplet, spray and combustion processes (Bisetti et al., 2012; Jenny et al., 2012; Pope, 2013; Shinjo and Umemura, 2013; Xia et al., 2013; Wang et al., 2014; Birouk and Toth, 2015; Kourmatzis and Masri, 2015), wind harvesting (Cao et al., 2011; Ahmadi-Baloutaki et al., 2015; Smith et al., 2015), and industrial particle production and transportation (Pratsinis and Srinivas, 1996; Fager et al., 2012; Sala and Marshall, 2013; Capecelatro et al., 2014), to name but a few applications, mostly sourced from a stockpile of recent publications. The erratic motion of fluid particles caused by fluctuating pressure gradients is fundamental to transport and mixing within a turbulent flow. But what is flow turbulence or turbulent flow? Although it

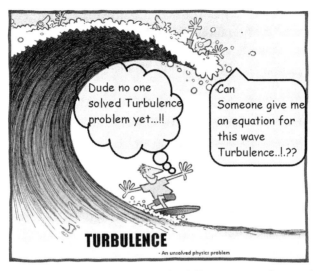

Figure 1.1 Turbulence, unresolved, and yet fearfully encountered every day. *(Created by S.P. Mupparapu).*

is challenging to answer this definitively, one of the most appealing descriptions is that "turbulence is the coexistence of structures and randomness." Before we go through some of the inherent characteristics associated with turbulent flows, the following quotations are worth appreciating.

Turbulence is the most important unsolved problem of classical physics.
– Richard P. Feynman (1918–1988), Nobelist Physicist

Figure 1.1 attempts to portray this trite but truthful statement.

It is trite to regard turbulence as the last unsolved problem in classical physics and to cite many books and authorities to justify the opinion. It is likewise cliché to list great physicists and mathematicians, such as Werner Heisenberg, Richard Feynman, and Andrei Kolmogorov, who "failed" to solve the problem despite much effort.
– G. Falkovich and K.R. Sreenivasan (2006)

This is not to say that there has not been any advancement. Maybe the beauty of flow turbulence is more to be contemplated than to be decoded.

For after all what is man in nature? A nothing in relation to infinity, all in relation to nothing, a central point between nothing and all and infinitely far from understanding either. The ends of things and their beginnings are impregnably concealed from him in an impenetrable secret. He is equally incapable of seeing the nothingness out of which he was drawn and the infinite in which he is engulfed.
– Blaise Pascal, Pensées #72

We are inclined to agree with Pascal, particularly when recalling a late eminent heat transfer professor who once stated that there is no end to research. This is definitely true in the case of engineering turbulence. The fact that turbulence research continues and has, in fact, increased in recent years indicates that our understanding of the topic has improved. Therefore, if we continue to make use of this existing research, we are in a position to provide substantial benefit to society. With this in mind, let us proceed to familiarize ourselves with the topic at hand.

1.1.1 Irregular or Random

First, it is important to recognize that turbulent flows are defined by their irregularity or randomness. Turbulent flows are unsteady and fluctuate randomly throughout the domains of space and time (Hinze, 1959).[1] In other words, turbulent flows have random velocity fluctuations with a wide range of length and time scales. This makes a deterministic approach to turbulence problems incredibly challenging. However, it welcomes the use of statistical methods to provide approximations and insights into certain problems. Thus, said Hinze,

> *Turbulent fluid motion is an irregular condition of flow in which the various quantities show a random variation with time and space coordinates, so that statistically distinct average values can be discerned.*
>
> *– Hinze (1959)*

Due to the unpredictable nature of turbulent flow, it should be noted that the word "discerned" should probably be replaced by "approximated." Nevertheless, flow turbulence has transport properties somewhat similar to those of molecular motion, but it is significantly more complex. This is because unlike molecular mass and mean free path in molecular motion, most characteristics of turbulent flow are not constants.

1.1.2 Rich in Scales of Eddying Motion

Second, turbulent flows are extremely rich in scales of eddying motion. As mentioned, a single turbulent flow will typically consist of a wide variation of length and time scales. These scales are the soul or the DNA of flow

---

[1] It is not sufficient to define turbulent motion as irregular in time alone. For example, an element of fluid within a volume, which is moving irregularly, is irregular with respect to time to a stationary observer, but not to an observer moving with the volume of fluid. Nor is turbulent motion a motion that is irregular in space alone, since a steady flow with an irregular flow pattern alone might then come under the definition of turbulence.

turbulence and thus are thoroughly expounded upon in Chapter 4. The large-scale motions are strongly influenced by the geometry of the flow, that is, boundary conditions, and they seem to control the transport and mixing within the flow. The behavior of the small-scale motions, on the other hand, may be determined almost entirely by the rate at which they receive energy from the large scales, although they are also influenced by the viscosity of the fluid. Therefore, these small-scale motions can have a universal character, independent of the flow geometry. The assumption of local isotropy has been challenged by some researchers, although when the conditions are right (e.g., high Re), this assumption appears to be quite reasonable. We note that our eyes see only a small window in time (and space) and are therefore likely to focus most heavily on the largest time/space scales in front of us. In other words, turbulence can be pictured as the superposition of eddies of ever-smaller sizes. These various-sized eddies have a certain amount of kinetic energy, as determined by their vorticity ($\Omega \equiv \nabla \times \mathbf{U}$, here $\mathbf{U}$ signifies the velocity vector) or by the intensity of the velocity fluctuation ($\frac{1}{2} mu^2$, where m is the mass and u is the fluctuating velocity) of the corresponding frequency. A distribution of energy, on average, between the frequencies is called an energy spectrum.

1.1.3 Large Reynolds Number

The other characteristic of turbulent flows is a large Reynolds number. Reynolds number is defined as: Re = inertia/viscous force = UL/v, where U is the velocity, L is a characteristic length, and v is the kinematic viscosity ($v = \mu/\rho$, where μ is the dynamic viscosity and ρ is the density). One can compare the flow of honey with that of a waterfall (e.g., Niagara Falls) as tabulated in Table 1.1 and depicted in Fig. 1.2. Honey has a large v and/or a small U and, therefore, a low inertia and a small Reynolds number. The large v removes fluctuations and dissipates the kinetic energy into heat. On the other hand, a waterfall has a relatively small v and/or a large U and, consequently, a high inertia and a large Reynolds number. Therefore, there is a lot of turbulence in a large waterfall; one can even detect the inertia from a

Table 1.1 Flow of honey versus that of a waterfall (Reynolds number illustrated)

Honey	Waterfall
Large v, small U	Small v, Large U
Small inertia	Large inertia
Small Re	Large Re

Figure 1.2 Flow of honey (small Re) versus that of a waterfall (large Re). *(Photos taken by Z. Ting and N. Ting).*

distance through vibrations in the ground and noise from the falling water. While not all turbulent flows are so majestic, they all have Re so large that fluid viscosity cannot keep the turbulence from occurring.

1.1.4 Dissipative

Turbulent flows are always dissipative in the sense that they lose energy and decay. Note that waves such as acoustic noise are dispersive but not dissipative; that is, they spread out without losing energy. Viscosity removes flow fluctuations (instabilities) by converting the associated kinetic energy into heat; this is the dissipation process. A flow remains laminar when small perturbations are damped out via viscosity. Viscous force becomes relatively small at larger Reynolds numbers, which means that this damping, derived from the molecular diffusion of momentum, is unable to dissipate the overwhelming perturbations (instabilities related to the interaction of viscous terms and nonlinear inertia terms in the equations of motion). As such, turbulence often originates from the instabilities in laminar flows when the Reynolds number becomes large. The perturbations can also originate from slight thermal current, surface roughness, etc., and perhaps even from microscopic sources such as the sub-continuum molecular motions that cause Brownian motion (Tsuge, 1974).

1.1.5 Highly Vortical

Turbulent flow is highly vortical, meaning that it is rotational and characterized by high levels of fluctuating vorticity. Note that although the characteristics of cyclones may be strongly influenced by the interaction of small-scale turbulence (generated by shear or buoyancy) with large-scale flow, they are not themselves turbulence. Likewise, random waves on the ocean's surface are not turbulence, as they are essentially irrotational.

The velocity derivatives are dominated by the smallest scales of turbulence. Vorticity is defined as the curl of the velocity, $\Omega \equiv \nabla \times \mathbf{U}$. Hence, vortex dynamics is a promising approach to studying and modeling turbulence. We will discuss this at length in Chapter 8.

1.1.6 Three-Dimensional

Turbulence is intrinsically three-dimensional. The term "two-dimensional turbulence" is only used to describe the simplified case where flow is restricted to two dimensions. Based on this description, we can note that two-dimensional turbulence is not true turbulence. Vorticity fluctuations cannot be two-dimensional because vortex stretching, an important vorticity-maintenance mechanism, is not present in a two-dimensional flow.

1.1.7 Highly Diffusive

Turbulent flows are also highly diffusive, and their diffusivity is much greater than that of a laminar flow (molecular diffusivity). The highly diffusive turbulence causes rapid mixing and increased rates of momentum, heat, and/or mass transfer. An easy example for students in a flow turbulence class to remember is also a smelly one. Suppose a student in the class ate too many beans or other healthy foods that promote flatulence and happened to release some of the by-product in the middle of the class. Most of the students, other than those seated next to the culprit, would probably not be aware of the release if the air in the classroom was largely stagnant. Under these conditions, the by-product could only spread via molecular diffusivity. In reality, thanks to turbulent air motion, everyone gets a dose of the gas within a minute or two. This dissipation is proof that the air in the room is turbulent. As such, even an apparently random flow pattern is not turbulent if it does not exhibit the spreading of velocity fluctuations throughout the surrounding fluid, as is the case with, for example, a constant diameter jet.

1.1.8 Turbulent Flows are Flows

Turbulence is not a feature of fluids but of fluid flows. Turbulence is different for different flows, even though all turbulent flows have many common characteristics. Thus, the research approach of borrowing from molecular diffusivity and/or gas kinetic theories and applying them to flow turbulence is fundamentally unfounded. Notwithstanding that, flows can be compared, provided they are not too dissimilar.

1.1.9 Continuum

Turbulence is a continuum phenomenon, governed by the equations of fluid mechanics. Even the smallest turbulent length scales are much larger than the molecular length. This is only a problem in abnormal molecular conditions, such as when dealing with very thin air or rarefied gas.

1.2 A BRIEF HISTORIC ACCOUNT

Turbulent flows are the norm in real life, while laminar flows are the oddities. We see turbulence in rivers, oceans, clouds, smoke, waterfalls, bloodstreams, etc. Such a common and yet perplexing phenomenon has infatuated many inquisitive minds for centuries. The following is a brief chronological account of some of the recorded attempts to decode turbulence.

1.2.1 Leonardo da Vinci (1452–1519)

It is generally agreed that Leonardo da Vinci, the renowned scientist, philosopher, and artist of the fifteenth century, was the first to tackle turbulence. Da Vinci's exceedingly detailed turbulence sketch reveals not only the master hand of a one-of-a-kind artist, but also the mind of a genius fascinated by flow turbulence.

1.2.2 Lord Rayleigh (1878, 1880)

Rayleigh conducted a series of theoretical investigations on the stability of a parallel flow of an inviscid fluid. He found that a necessary and sufficient condition for a parallel-flow inviscid fluid to become unstable is for it to possess an inflection point. An inflection point signifies that some sort of flow deceleration is required for an inviscid flow to become unstable.

We discussed earlier that the flow of a real (viscous) fluid will only become turbulent when the inertia is larger than the viscosity. In other words, viscosity damps out turbulence. Rayleigh's discovery, however, suggests that the presence of viscosity can promote the initiation of turbulence by enabling the creation of inflection points. Therefore, viscosity is a double-edged sword – it is needed to kill turbulence, but it can also simultaneously create it.

1.2.3 Osborne Reynolds (1842–1912)

In 1883, Osborne Reynolds built 6-ft. long glass tubes with diameters of 2.68, 1.53, and 0.789 cm with trumpet mouths and passed water through them (Reynolds, 1883, 1894). He found that at a low flow rate and/or with a small diameter tube, an injected color streak is seen as a steady streak (Fig. 1.3a). In other words, the flow is laminar at low Re where small perturbations are damped out by viscosity. Note that a laminar flow in a smooth long pipe in Fig. 1.3a is named after Poiseuille.[2]

At a higher flow rate, the color band shown in Fig. 1.3b appears to expand and mix with the water. When viewing the tube by the light of an electric spark (Fig. 1.3c), the mass of color resolves itself into a mass of more or less distinct curls, in which eddies can be seen.

In the transitional realm of an intermediate flow rate, as depicted by Fig. 1.3d, we see the intermittent character of the flow motion caused by the disturbances, which appear as flashes succeeding each other inside the tube. In other words, we see sporadic bursts of turbulence alternating with laminar flow, indicating a problem with multiple solutions. Consequently, some have approached the origin of turbulence via the bifurcation/chaos method.

---

[2] Poiseuille flow is a mathematical possibility in an infinite pipe since it is an exact solution of the Navier–Stokes equations. Poiseuille flow is stable to infinitesimal perturbations at all Re. This implies that transition to turbulence is dependent on the perturbation, in addition to Re. This is unlike flows around a bluff body where instability can be predicted to occur using linearized perturbation theory with an infinitesimal amplitude. The above two points indicate that there is more than one way to create a turbulent flow.

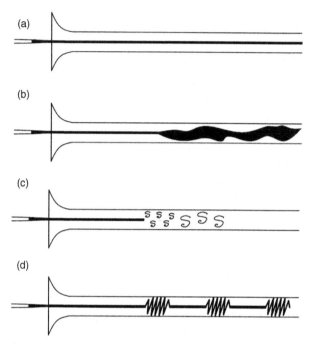

Figure 1.3 Osborne Reynolds' experiment on laminar to turbulent flow in a pipe. *(Created by H. Cen).*

For practical pipe flows, transition typically occurs at $Re \approx 2000$ in the form of decaying turbulent slugs passing through the pipe, and a fully turbulent state generally occurs when $Re > 2300$. It is, however, not possible to be precise about these critical Reynolds numbers, as they are very sensitive to upstream flow conditions and the texture (smoothness or roughness) of the boundary. With sufficient care taken to minimize possible disturbances, laminar flow in a pipe can be maintained to at least $Re \approx 10^5$ (Pfenninger, 1961).

An idealized sketch of the laminar–turbulent transition process over a flat plate is depicted in Fig. 1.4. We see that the vortical structures are initially two-dimensional in nature, generated due to shear (no-slip conditions) in the boundary layer. These vortical structures become progressively unstable farther downstream, with three-dimensional interactions eventually leading to turbulence.

1.2.4 Henri Bénard (1900)

Consider a fluid confined between two large parallel plates. When the temperature of the lower plate is marginally higher than that of the top plate,

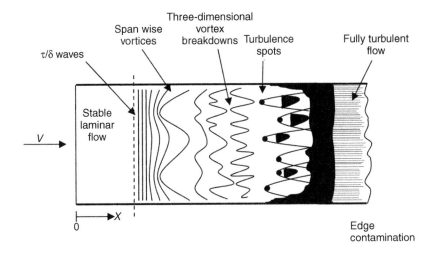

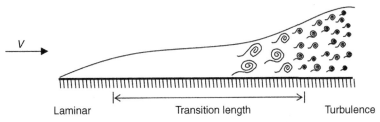

Figure 1.4 Laminar–turbulent transition process on a flat plate. *(Created by A. Vasel-Be-Hagh).*

heat is conducted through the stagnant (quiescent) fluid via molecular conduction. The upward buoyancy force is balanced by the vertical pressure gradient until a critical temperature difference is reached and laminar Bénard cells are formed (see Fig. 1.5). It is found that the instability criterion is

$$\mathrm{Ra} \equiv \text{gravity/thermal diffusivity} = g\beta\Delta Th^3/(v\alpha) = 1.70 \times 10^3 \quad (1.1)$$

where β is the expansion coefficient, h is the height of the gap, and α is the thermal diffusivity. Note that this critical Rayleigh number (Ra) coincides almost exactly (off by only 0.7%) with the critical Taylor number (which we will cover next), and the wave number of the convection cells is $3.12/h$.

Further increase in Ra will eventually lead to the Bénard cells themselves becoming unstable. In a case of fully turbulent convection, the flow field consists of a time-averaged component plus a random, chaotic motion.

Cold

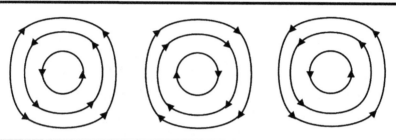

Hot

Figure 1.5 Bénard convection cells. *(Created by A. Vasel-Be-Hagh).*

1.2.5 Taylor (1915, 1923, 1935, 1938)

Taylor studied the flow between two concentric cylinders, with the inner cylinder rotating at a constant speed. At low rotation rates, the fluid within the gap, being dragged around by the inner cylinder, also rotates. At a certain critical speed, toroidal (Taylor) vortices appear due to instability of the basic rotary flow, superimposed on the primary circular motion (see Fig. 1.6a). When the cylinders are very long and the gap is very narrow, only the Taylor number (Ta) determines the onset of the Taylor vortices.

The formation of Taylor vortices has to do with the centrifugal force, which tends to drive the rotating fluid radially outward. Below the critical speed, this force is balanced by the pressure gradient and the viscous force. Above the critical speed, the prevailing centrifugal force drives the rotating fluid outward. Because the outer fluid is in the way, the outwardly moving fluid breaks up into cells as shown in Fig. 1.6a. This is known as Rayleigh instability.[3]

At higher rotational speed, the Taylor vortices themselves become unstable and *wavy Taylor vortices* appear (see Fig. 1.6b). Even though the structures are more complex, this flow is still laminar.

Modulated wavy Taylor vortices emerge with a further increase in speed. When the speed is sufficiently high, the flow becomes fully turbulent (see Fig. 1.6c). At this point, the time-averaged flow pattern resembles that of the steady Taylor vortices, only the cells are somewhat larger (compare Fig. 1.6c with Fig. 1.6a).

---

[3] Rayleigh identified the instability mechanism and produced a stability criterion for inviscid, rotating flows. Taylor later extended the theory to viscous flows.

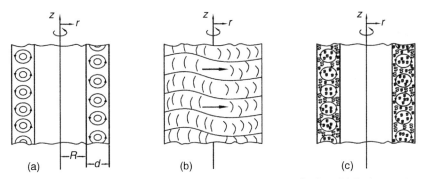

Figure 1.6 Taylor vortices for flow between two concentric cylinders. (a) Taylor vortices; (b) wavy Taylor vortices; and (c) turbulent Taylor vortices. *(Created by H. Cen).*

The three independent dimensionless groups are Taylor number, $\text{Ta} \equiv$ centrifugal force/viscous force $= \Omega^2\, s^3\, r_{inner}/v^2$; s/r_{inner}; and L/r_{inner}, where Ω is the rotation rate, s is the gap (spacing), r_{inner} is the radius of the inner cylinder, and L is the length of the apparatus. When $L \gg r_{inner}$ and $s \ll r_{inner}$, the flow is completely characterized by Ta alone. In this case, $\text{Ta} \approx 1.70 \times 10^3$ and the axial wave number of the vortices is $2\pi/\lambda$, where λ is the wavelength.

Also, Taylor borrowed ideas from the kinetic theory of gases for treating homogeneous and isotropic turbulent flows, replacing the fluid viscosity with eddy viscosity. He used the term "turbulence spectrum" to describe turbulence, stating that turbulence at any point may be considered as an infinite sum of harmonic components, each having a unique scale (eddy size). Taylor also introduced velocity correlations for describing turbulent structure using the degree with which velocity components at neighboring points are correlated. The statistical approach to understanding flow turbulence was also more or less initiated by Taylor.

1.2.6 Prandtl (1925)

In line with the mean free path in the kinetic theory of gases, Prandtl introduced the concept of a "mixing length," wherein the mixing length is the average distance a fluid element would stray from the mean streamline. Recall that a streamline is a line that is always tangent to the velocity vector at a given instant. As mentioned before, turbulence is not a feature of fluids but one of fluid flows. Thus, the analogy between "mean free path" and "mixing length," though useful in many aspects, is fundamentally invalid.

The above historic review is far from complete. Interested readers are referred to specialized documents such as Laufer (1975), Townsend (1990), Lumley and Yaglom (2001). The manifold contributions of Taylor and Prandtl will be expounded upon in later chapters.

1.3 ORGANIZATION OF THE BOOK

Chapter 2 details the basic equations of fluid in motion. This is followed by the promising statistical description of flow turbulence in Chapter 3. The crux of the book is turbulence scales, and thus, Chapter 4 is comprehensively devoted to this topic. Chapter 5 briefly highlights turbulence simulations and modeling. Wall turbulence is reviewed in Chapter 6. The practically simplest and cleanest grid turbulence is the subject of Chapter 7. Chapter 8 talks about vortex dynamics in the context of understanding and modeling flow turbulence. The last two chapters provide common topics of engineering problems involving turbulence. Chapter 9 discusses a sphere and a circular cylinder in cross flow, while Chapter 10 discusses premixed turbulent flow propagation. We hope that inquisitive scholars who journey on this turbulence path will fall in love with flow turbulence along the way, whether voluntarily or not (see Fig. 1.7).

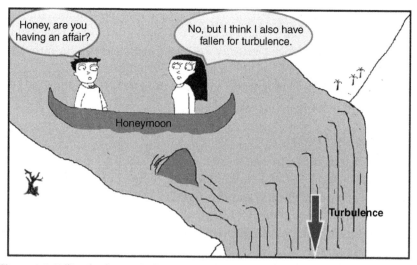

Figure 1.7 Falling for flow turbulence. *(Created by H. Wu and edited by D. Ting).*

Problems

Problem 1.1 Dimensional analysis

A turbulent flow researcher aims at studying the wake behind two identical spheres falling above the free surface of water. The resulting wake is thought to depend on the sphere diameter, the distance between the two spheres, the initial height from the free surface, surface tension, viscosity, distance or time traveled, etc. How many dimension-less parameters are required to characterize this? What are the parameters?

Problem 1.2 Turbulence or not?

Give a concise and clear example, with brief explanation, of a turbulent flow (or flow turbulence). Give another example of a flow that seems turbulent but is not.

Problem 1.3 Turbulence from a waterfall

A stream of waterfalls down into a deep lake via a height of 7 m at a rate of 25 kg/s. Have a rough estimate of the integral scale of the turbulence. Also estimate the size of the smallest eddies.

REFERENCES

Ahmadi-Baloutaki, M., Carriveau, R., Ting, D.S.-K., 2015. An experimental study on the interaction between free-stream turbulence and a wing-tip vortex in the near-field. Aerosp. Sci. Technol. 43, 395–405.

Birouk, M., Toth, S.L., 2015. Vaporization and combustion of a soybean biodiesel droplet in turbulent environment at elevated ambient pressure. Combust. Sci. Technol. 187, 937–952.

Bisetti, F., Blanquart, G., Mueller, M.E., Pitsch, H., 2012. On the formation and early evolution of soot in turbulent nonpremixed flames. Combust. Flame 159, 317–335.

Cao, N., Ting, D.S.-K., Carriveau, R., 2011. The performance of a high-lift airfoil in turbulent wind. Wind Eng. 35 (2), 179–196.

Capecelatro, J., Desjardins, O., Fox, R.O., 2014. Numerical study of collisional particle dynamics in cluster-induced turbulence. J. Fluid. Mech. 747 (R2), 1–13.

Fager, A.J., Liu, J., Garrick, S.C., 2012. Hybrid simulations of metal particle nucleation: a priori and a posteriori analyses of the effects of unresolved scalar interactions on nanoparticle nucleation. Phys. Fluids 24 (075110), 1–18.

Falkovich, G., Sreenivasan, K.R., 2006. Physics Today, April, pp.43–49.

Grabowski, W.W., Wang, L.-P., 2013. Growth of cloud droplets in a turbulent environment. Annu. Rev. Fluid Mech. 45, 293–324.

Hinze, J.O., 1959. Turbulence. McGraw-Hill, USA.

Jenny, P., Roekaerts, D., Beishuizen, N., 2012. Modeling of turbulent dilute spray combustion. Prog. Energy Combust. Sci. 38, 846–887.

Kourmatzis, A., Masri, A.R., 2015. Air-assisted atomization of liquid jets in varying levels of turbulence. J. Fluid Mech. 764, 95–132.

Laufer, J., 1975. New trends in experimental turbulence research. Annu. Rev. Fluid Mech. 7, 307–326.

Lumley, J.L., Yaglom, A.M., 2001. A century of turbulence. Flow, Turbulence and Combustion. 66, 241–286.

Pfenninger, W., 1961. Boundary layer suction experiments with laminar flow at high Reynolds numbers in the inlet length of tubes of various suction methods. In: Lachmann, G.V. (Ed.), Boundary Layer and Flow Control. Pergamon Press, Oxford, pp. 961–980.

Pope, S.B., 2013. Small scales, many species and the manifold challenges of turbulent combustion. P. Combust. Inst. 34, 1–31.

Prandtl, L., 1925. Bericht über untersuchingen zur ausgebildeten turbulenz. Z. Angew. Math. Mech. 5, 136–137.

Pratsinis, S.E., Srinivas, V., 1996. Particle formation in gases, a review. Powder Technol. 88, 267–273.

Rayleigh, L., 1878. On the instability of jets. Proc. Lond. Math. Soc. 10, 4–13. Scientific Papers 1, 361–371. Cambridge University Press.

Rayleigh, L., 1880. On the stability, or instability, of certain fluid motions. Proc. Lond. Math. Soc. 11, 57–70. Scientific Papers 1, 474–487. Cambridge University Press.

Reynolds, O., 1883. An experimental investigation of the circumstances which determine whether the motion of water shall be direct or sinuous, and of the law of resistance in parallel channels. Philos. Trans. R. Soc. A 174, 935–982.

Reynolds, O., 1894. On the dynamical theory of incompressible viscous fluids and the determination of the criterion. Philos. Trans. R. Soc. A 186, 123–161.

Sala, K., Marshall, J.S., 2013. Stochastic vortex structure method for modeling particle clustering and collisions in homogeneous turbulence. Phys. Fluids 25 (103301), 1–21.

Shinjo, J., Umemura, A., 2013. Droplet/turbulence interaction and early flame kernel development in an autoigniting realistic dense spray. Proc. Combust. Inst. 34, 1553–1560.

Smith, J., Carriveau, R., Ting, D.S.-K., 2015. Turbine power production sensitivity to coastal sheared and turbulent inflows. Wind Eng. 39 (2), 183–192.

Taylor, G.I., 1915. Eddy motion in the atmosphere. Philos. Trans. R. Soc. Lond. A215, 1–26.

Taylor, G.I., 1923. Diffusion by continuous movements. Proc. Lond. Math. Soc. 20 (2), 196–211.

Taylor, G.I., 1935. Statistical theory of turbulence. Parts 1–4. Proc. R. Soc. Lond. A A151, 421–478.

Taylor, G.I., 1938. The spectrum of turbulence. Proc. R. Soc. Lond. A A164, 476–479.

Townsend, A.A., 1990. Early days of turbulence research at Cambridge. J. Fluid Mech. 212, 1–5.

Tsuge, S., 1974. Approach to origin of turbulence on basis of 2-point kinetic theory. Phys. Fluids 17, 22–33.

Vaillancourt, P.A., Yau, M.K., 2000. Review of particle-turbulence interactions and consequences for cloud physics. Bull. Am. Meteorol. Soc. 81, 285–298.

Wang, J.H., Wei, Z.L., Zhang, M., Huang, Z.H., 2014. A review of engine application and fundamental study on turbulent premixed combustion of hydrogen enriched natural gas. Sci. China Ser. E 57 (3), 445–451.

Weil, J.C., Sykes, R.I., Venkatram, A., 1992. Evaluating air-quality models: review and outlook. J. Appl. Meteorol. 31, 1121–1145.

Xia, J., Zhao, H., Megaritis, A., Luo, K.H., Cairns, A., Ganippa, L.C., 2013. Inert-droplet and combustion effects on turbulence in a diluted diffusion flame. Combust. Flame 160, 366–383.

CHAPTER 2

Equations of Fluid in Motion

The future belongs to those who believe in the beauty of their dreams.
– Eleanor Roosevelt

Contents

Chapter Objectives

- To understand the continuum approach.
- To discern and differentiate the Lagrangian framework from the Eulerian framework.
- To learn about the Eulerian-Lagrangian transformation.
- To derive the equations of motion for laminar flow.
- To understand Reynolds decomposition.
- To extend the equations of motion to turbulent flow via Reynolds decomposition.

NOMENCLATURE

A	Area
CAD	Crank angle degree
F	Force
k	Thermal conductivity
Kn	Knudsen number, molecular mean free path/characteristic physical length
l	Smallest geometric length in a flow
l_{medium}	Intermediate length scales
m	Mass

Basics of Engineering Turbulence
http://dx.doi.org/10.1016/B978-0-12-803970-0.00002-7

N	Total number of cycles
n	Cycle number
P	Pressure
p	Fluctuating pressure
STP	Standard Temperature and Pressure (25°C and 1 atm)
T	Temperature
T_{period}	Time period
t	Time
U	Velocity, or velocity in the x direction
u	Fluctuating velocity in the x direction
V	Velocity in the y direction, volume
v	Fluctuating velocity in the y direction
W	Velocity in the z direction
w	Fluctuating velocity in the z direction
x, y, z	Cartesian coordinates
Bold	$\Rightarrow$ Tensor
Overbar	$\Rightarrow$ Time-averaged or mean

Greek symbols

α	Field variable
γ	Angle
δ	A small amount
λ_{mfp}	Molecular mean free path
μ	Dynamic (absolute) viscosity
v	Kinematic viscosity
ρ	Density
σ	Stress
τ	Shear; timescale

2.1 INTRODUCTION

In this chapter, we will invoke basic principles and derive the fundamental equations of fluid in motion. To do so we first introduce the continuum concept, followed by the Eulerian and Lagrangian frameworks and the transformation from one frame to the other via the Reynolds' transport theorem. The equations corresponding to the familiar laminar case are derived before applying Reynolds decomposition to deduce those for the turbulent case.

The conservation equations for fluid can be derived in two basic ways, either statistically or under the assumption of continuum mechanics. The statistical approach tackles the problem from a molecular point of view. It treats the fluid as a group of molecules where motion is governed by the laws of dynamics. It predicts macroscopic behavior from the laws of mechanics and probability theory. The transport coefficients such as the (kinematic) viscosity ν and the thermal conductivity k are functions of molecular forces. As such, the statistical approach works well for light gases where the molecules are relatively sparse. It is, however, incomplete for polyatomic gas molecules and for liquids.

The continuum approach assumes the fluid to consist of continuous matter, rather than discrete particles. At each point of this continuous fluid, there is supposed to be a unique value of the velocity, pressure, density, and other field variables. The continuous matter obeys the laws of conservation. This leads to a set of differential equations governing the field variables, the solution of which defines the variation of each field variable with respect to space and time. At any instant in time, the variable assumes the mean value of the molecular magnitude at that location. For the continuum approach to be valid, the mean free path of the molecules must be very small relative to the smallest physical length scale. In other words, the continuum hypothesis is good when the Knudsen number, the ratio of the molecular mean free path length to the representative physical length scale, is small. Specifically

$$\text{Kn}(\equiv \lambda_{mfp} \, / \, l) << 1 \tag{2.1}$$

where λ_{mfp} is the molecular mean free path (for air, this is approximately 6×10^{-8} m at STP [standard temperature and pressure]) and l is the smallest geometric length scale in a flow.

For a very small Kn, there exists a relatively intermediate length l_{medium} which is large compared to λ_{mfp}, but small compared to l. Under this condition, the continuum fluid properties such as density and velocity can be thought of as the molecular properties averaged over a volume of size $(l_{medium})^3$. This is utilized in the derivation of differential conservation equations, where a control volume is shrunk to infinitesimal size. In Cartesian coordinates, this infinitesimal volume has dimensions dx, dy, and dz, giving a volume of $dx \, dy \, dz$, which is equivalent to $(l_{medium})^3$, as the derivation invokes the continuum assumption.

Beyond a certain distance away from the earth surface, say, at an elevation of 100 km, atmospheric air has a mean free path on the order of 0.1 m. It is therefore not surprising to see the continuum approach break down and the aforementioned statistical approach (the rarefied gas flow theory) become more appropriate. We shall stay with the continuum approach throughout this book; that is, we only deal with conditions where it is valid. Unless otherwise stated, we will further limit ourselves to incompressible Newtonian fluids only.

When dealing with very high levels of flow turbulence, the size of the eddying motion can become quite small. Even in typical high-intensity turbulent flows, the length scale of the smallest eddies is generally still much larger than typical intermolecular distances; hence, the fluid may be modeled as a **continuum** medium. This book only considers situations where this holds true.

2.2 EULERIAN AND LAGRANGIAN FRAMES

The two choices of reference frameworks are the Eulerian and the Lagrangian. Reynolds' transport theorem, to be covered later, can be used to relate derivatives in these two frameworks. In other words, the derived equations can be transformed from one framework to the other via the Reynolds' transport theorem (Currie, 1974).

2.2.1 Eulerian

Eulerian fields are indexed by the position vector x in an inertial frame, usually by fixing a control volume in space and monitoring the flow passing through the control volume over a period of time. As such, the independent variables are the spatial coordinates (x, y, z) and the time (t). In other words, the properties of a flow field are specified in terms of space coordinates and time. This approach is typically preferred for solving fluid dynamics problems.

2.2.2 Lagrangian

The Lagrangian approach is gaining ground in the detection of Lagrangian coherent structures, which are free from the uncertainties associated with single trajectories (Haller, 2015). The full Lagrangian skeleton of material surfaces, the Lagrangian coherent structures, of a general turbulent flow such as that in an ocean or the atmosphere can now be determined (Haller, 2015). This Lagrangian approach that keeps track of a particular mass of fluid has also been traditionally used to derive the basic equations. Consider the generic case where the fluid particle initially (at time t_0) is at position $X^+(t_0, Y) = Y$, where Y is the Lagrangian or material coordinate. At any time t, the particle is at spatial location X^+, and at that position the particle moves with the local fluid velocity; that is

$$\frac{\partial \vec{X}^+\left(t, \vec{Y}\right)}{\partial t} = \vec{U}\left(\vec{X}^+\left(t, \vec{Y}\right), t\right) \qquad (2.2)$$

Note that the local fluid velocity can be deduced from the Eulerian velocity field $\mathbf{U}(\mathbf{x}, t)$. Hence, knowing the initial particle position and the Eulerian velocity field, we can trace the path of the particle under consideration.

2.2.3 Eulerian-Lagrangian Transformation

Let α be any field variable (concentration, ρ, T) of the fluid. From the Eulerian point of view, α may be considered to be a function of the independent variables x, y, z, and t. At any spatial position (x, y, z), the value of α is only a function of time t; and it is a constant under steady flow conditions. Alternatively speaking, at any time instant t or under steady flow condition, the value of α is specified explicitly by the spatial coordinates (x, y, z).

On the other hand, if we follow a specific fluid element over a short period of time δt, its position will change by amounts $\delta x, \delta y$, and δz, while the value of α will change by an amount $\delta \alpha$. In this Lagrangian framework, the independent variables are x_0, y_0, z_0, and t, where x_0, y_0, and z_0 are the initial coordinates for the fluid element. As such, x, y, and z are no longer independent variables as in the Eulerian framework, but are functions of t as defined by the trajectory of the element. Over the short time period δt, the change in α may be deduced from the differential calculus

$$\alpha\left(t, x, y, z\right) \Rightarrow \frac{\partial \alpha}{\partial t}\delta t + \frac{\partial \alpha}{\partial x}\delta x + \frac{\partial \alpha}{\partial y}\delta y + \frac{\partial \alpha}{\partial z}\delta z \tag{2.3}$$

Equating the above change in α to the observed change $\delta \alpha$ in the Lagrangian framework, we have

$$\delta \alpha)_{t \to t+\delta t} = \frac{\partial \alpha}{\partial t}\delta t + \frac{\partial \alpha}{\partial x}\delta x + \frac{\partial \alpha}{\partial y}\delta y + \frac{\partial \alpha}{\partial z}\delta z \tag{2.4}$$

Dividing through by δt gives

$$\frac{\delta \alpha}{\delta t} = \frac{\partial \alpha}{\partial t} + \frac{\delta x}{\delta t}\frac{\partial \alpha}{\partial x} + \frac{\delta y}{\delta t}\frac{\partial \alpha}{\partial y} + \frac{\delta z}{\delta t}\frac{\partial \alpha}{\partial z} \tag{2.5}$$

The left-hand side represents the total change in α over a brief time interval δt in the Lagrangian framework. In the limit where the time interval is indefinitely small, the left-hand side represents the time derivative of α in the Lagrangian system, $D\alpha/Dt$, which is called the material derivative. In other words, as δt approaches zero, we have

$$\frac{D\alpha}{Dt} = \frac{\partial \alpha}{\partial t} + U\frac{\partial \alpha}{\partial x} + V\frac{\partial \alpha}{\partial y} + W\frac{\partial \alpha}{\partial z} \tag{2.6}$$

The first term on the right-hand side depicts the change in α with respect to time, and thus it is zero for the steady case. The last three terms on the right-hand side signify the change in α due to convection; the first represents the change in α when the fluid element is convected by the velocity in the x direction, the second is the change due to convection in the y direction, and the third is the change caused by convection in the z direction. The equation can be expressed in vector form

$$\frac{D\alpha}{Dt} = \frac{\partial \alpha}{\partial t} + \left(\vec{U} \cdot \nabla\right)\alpha \tag{2.7}$$

Alternatively, we can express it using Einstein's summation convention

$$\frac{D\alpha}{Dt} = \frac{\partial \alpha}{\partial t} + U_k\frac{\partial \alpha}{\partial x_k} \tag{2.8}$$

As mentioned earlier, the entire right-hand side represents the total change in α expressed in Eulerian coordinates. The first term on the right-hand side shows that at any point in space, the fluid properties may change with respect to time; this is the unsteady scenario. In a time–independent, steady flow field, the second term illustrates that the fluid property α changes with respect to the location only. In the general case, α can vary with respect to both time and space.

It is worth emphasizing that $D\alpha/Dt$, which represents the total change in the quantity α as seen by an observer following a fluid element, is called the **material derivative** or **convective derivative**. The definition is believed to have been first introduced by Stokes.

2.3 COMMON EQUATIONS IN FLUID MECHANICS

In this section, the conservation of mass, momentum, and energy are derived for a flow without turbulent fluctuations. These derivations are available in standard fluid mechanics textbooks such as Fox et al. (2009). They

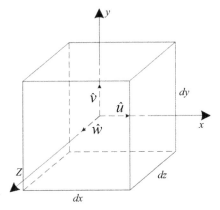

Figure 2.1 A differential control volume in the Cartesian coordinate. *(Created by Z. Yang).*

are provided here in detail to illustrate the additional terms associated with fluctuating turbulent flows in Section 2.5.

2.3.1 Conservation of Mass

In the absence of nuclear reactions and other extreme conditions such as those involving a mass traveling at the speed of light, mass can neither be destroyed nor created. This conservation of mass principle is also referred to as the continuity principle. Consider an infinitesimal control volume of dimensions $dx\,dy\,dz$, as shown in Fig. 2.1. At the center of the volume, point 0 with coordinates $(0, 0, 0)$, the fluid density is ρ and the velocity is

$$\vec{U} = U\hat{i} + V\hat{j} + W\hat{k} \tag{2.9}$$

The pointed hats are used to denote unit vectors. Invoking Taylor series expansion about point 0 leads to terms such as

$$\rho\big)_{x+\frac{dx}{2}} = \rho + \frac{\partial \rho}{\partial x}\frac{dx}{2} + \frac{\partial^2 \rho}{\partial x^2}\frac{1}{2!}\left(\frac{dx}{2}\right)^2 + \cdots \tag{2.10}$$

Neglecting the much smaller, higher order terms, we are left with

$$\rho\big)_{x+\frac{dx}{2}} = \rho + \left(\frac{\partial \rho}{\partial x}\right)\frac{dx}{2} \tag{2.11}$$

$$U\big)_{x+\frac{dx}{2}} = U + \left(\frac{\partial U}{\partial x}\right)\frac{dx}{2} \tag{2.12}$$

where ρ, U, $\partial \rho/\partial x$, $\partial U/\partial x$ are evaluated at point 0.

The conservation of mass for the control volume $dx\,dy\,dz$ depicted in Fig. 2.1 implies that the rate of mass entering the control volume minus that exiting the control volume is equal to the rate of change of mass of the control volume (element). In other words

$$\dot{m}_{in} - \dot{m}_{out} = \frac{\partial}{\partial t} m_{element} \qquad (2.13)$$

The mass flux through each of the six surfaces of the control volume shown in Fig. 2.1 can be described as

$$\int_{CS} \rho \vec{U} \cdot d\vec{A} \qquad (2.14)$$

Here, subscript "CS" signifies the control surface, and **A** (or A with an arrow head) denotes the surface tensor.

For the left ($-x$) surface, we have

$$-\left(\rho - \frac{\partial \rho}{\partial x}\frac{dx}{2}\right)\left(U - \frac{\partial U}{\partial x}\frac{dx}{2}\right)dy\,dz = -\rho U\,dy\,dz + \frac{1}{2}U\frac{\partial \rho}{\partial x}dx\,dy\,dz$$
$$+ \frac{1}{2}\rho\frac{\partial U}{\partial x}dx\,dy\,dz \qquad (2.15)$$
$$- \frac{1}{4}\frac{\partial \rho}{\partial x}\frac{\partial U}{\partial x}(dx)^2\,dy\,dz$$

Dropping the much smaller, higher (4th) order term leaves us with

$$-\left(\rho - \frac{\partial \rho}{\partial x}\frac{dx}{2}\right)\left(U - \frac{\partial U}{\partial x}\frac{dx}{2}\right)dy\,dz = -\rho U\,dy\,dz$$
$$+ \frac{1}{2}\left(U\frac{\partial \rho}{\partial x} + \rho\frac{\partial U}{\partial x}\right)dx\,dy\,dz \qquad (2.16)$$

Similarly, for the right ($+x$) surface, we have

$$\left(\rho + \frac{\partial \rho}{\partial x}\frac{dx}{2}\right)\left(U + \frac{\partial U}{\partial x}\frac{dx}{2}\right)dy\,dz = \rho U\,dy\,dz$$
$$+ \frac{1}{2}\left(U\frac{\partial \rho}{\partial x} + \rho\frac{\partial U}{\partial x}\right)dx\,dy\,dz \qquad (2.17)$$

The expression for the bottom $(-y)$ surface is

$$-\left(\rho - \frac{\partial \rho}{\partial y}\frac{dy}{2}\right)\left(V - \frac{\partial V}{\partial y}\frac{dy}{2}\right)dx\,dz = -\rho V\,dx\,dz$$
$$+\frac{1}{2}\left(V\frac{\partial \rho}{\partial y} + \rho\frac{\partial V}{\partial y}\right)dx\,dy\,dz \tag{2.18}$$

The flux entering the top $(+y)$ surface can be described by

$$\left(\rho + \frac{\partial \rho}{\partial y}\frac{dy}{2}\right)\left(V + \frac{\partial V}{\partial y}\frac{dy}{2}\right)dx\,dz = \rho V\,dx\,dz$$
$$+\frac{1}{2}\left(V\frac{\partial \rho}{\partial y} + \rho\frac{\partial V}{\partial y}\right)dx\,dy\,dz \tag{2.19}$$

And for the back $(-z)$ surface

$$-\left(\rho - \frac{\partial \rho}{\partial z}\frac{dz}{2}\right)\left(W - \frac{\partial W}{\partial z}\frac{dz}{2}\right)dx\,dy = -\rho W\,dx\,dy$$
$$+\frac{1}{2}\left(W\frac{\partial \rho}{\partial z} + \rho\frac{\partial W}{\partial z}\right)dx\,dy\,dz \tag{2.20}$$

Similarly, for the front $(+z)$ surface

$$\left(\rho + \frac{\partial \rho}{\partial z}\frac{dz}{2}\right)\left(W + \frac{\partial W}{\partial z}\frac{dz}{2}\right)dx\,dy = \rho W\,dx\,dy$$
$$+\frac{1}{2}\left(W\frac{\partial \rho}{\partial z} + \rho\frac{\partial W}{\partial z}\right)dx\,dy\,dz \tag{2.21}$$

Summing all the terms associated with the six surfaces gives

$$-\left[\frac{\partial \rho U}{\partial x} + \frac{\partial \rho V}{\partial y} + \frac{\partial \rho W}{\partial z}\right]dx\,dy\,dz = \frac{\partial \rho}{\partial t}\,dx\,dy\,dz \tag{2.22}$$

We can bring the right-hand term to the left and get

$$\frac{\partial \rho}{\partial t} + \frac{\partial \rho U}{\partial x} + \frac{\partial \rho V}{\partial y} + \frac{\partial \rho W}{\partial z} = 0 \tag{2.23}$$

The first term shows that local changes to mass (per unit volume) can occur when there is a change in the fluid density or when the fluid is compressible.

The remaining three terms signify mass changes (per unit volume) resulting from convection.

Alternatively, we see that the net rate of mass flux through the control surface is

$$\left(\frac{\partial \rho U}{\partial x} + \frac{\partial \rho V}{\partial y} + \frac{\partial \rho W}{\partial z} \right) dx\,dy\,dz \qquad (2.24)$$

and the rate of change of mass inside the control volume is

$$\frac{\partial \rho}{\partial t} dx\,dy\,dz \qquad (2.25)$$

therefore, the net rate of change of mass is

$$\frac{\partial \rho}{\partial t} dx\,dy\,dz + \left(\frac{\partial \rho U}{\partial x} + \frac{\partial \rho V}{\partial y} + \frac{\partial \rho W}{\partial z} \right) dx\,dy\,dz = 0 \qquad (2.26)$$

Dividing both sides by the volume $dx\,dy\,dz$ gives Eq. 2.23, which is the mass conservation or continuity equation with a general expression

$$\frac{\partial \rho}{\partial t} + \nabla \cdot \left(\rho \vec{U} \right) = 0 \qquad (2.27)$$

In the special case of incompressible fluid or constant-density flow, the above continuity equation is reduced to

$$\nabla \cdot \mathbf{U} = 0 \qquad (2.28)$$

This expression states that the total convection of mass into the control volume minus that convected out of the control volume is zero for a constant-density flow.

2.3.2 Momentum Equation

Applying Newton's second law to an infinitesimal fluid element of dimensions $dx\,dy\,dz$ can provide a dynamic equation describing the corresponding motion of the fluid element; see, for example, Çengel and Cimbala (2013) and Fox et al. (2009). According to Newton's second law, force is equal to mass times acceleration, and when applied to an infinitesimal system of mass dm, we have

$$d\vec{F} = dm \frac{d\vec{U}}{dt} \bigg)_{system} \qquad (2.29)$$

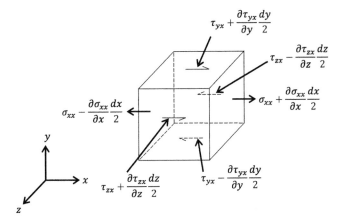

Figure 2.2 Laminar stresses on a control volume in the Cartesian coordinate. *(Created by P.K. Pradip).*

This can be rewritten as

$$dF = dm \frac{D\vec{U}}{Dt} = dm \left[U \frac{\partial \vec{U}}{\partial x} + V \frac{\partial \vec{U}}{\partial y} + W \frac{\partial \vec{U}}{\partial z} + \frac{\partial \vec{U}}{\partial t} \right] \quad (2.30)$$

Consider only the stresses that act in the x direction. The stresses at the center of the differential element are σ_{xx}, τ_{yx}, τ_{zx}, and the stresses acting in the x direction on all faces of the element are as shown in Fig. 2.2.

We can sum up all the pertinent forces to get the net surface force in the x direction

$$dF_{S_x} = \left(\sigma_{xx} + \frac{\partial \sigma_{xx}}{\partial x} \frac{dx}{2} \right) dy\, dz - \left(\sigma_{xx} - \frac{\partial \sigma_{xx}}{\partial x} \frac{dx}{2} \right) dy\, dz + \left(\tau_{yx} + \frac{\partial \tau_{yx}}{\partial y} \frac{dy}{2} \right) dx\, dz$$

$$- \left(\tau_{yx} - \frac{\partial \tau_{yx}}{\partial y} \frac{dy}{2} \right) dx\, dz + \left(\tau_{zx} + \frac{\partial \tau_{zx}}{\partial z} \frac{dz}{2} \right) dx\, dy - \left(\tau_{zx} - \frac{\partial \tau_{zx}}{\partial z} \frac{dz}{2} \right) dx\, dy$$

$$(2.31)$$

Canceling out the equivalent terms with opposing signs leaves us with

$$dF_{S_x} = \left(\frac{\partial \sigma_{xx}}{\partial x} + \frac{\partial \tau_{yx}}{\partial y} + \frac{\partial \tau_{zx}}{\partial z} \right) dx\, dy\, dz \quad (2.32)$$

With gravity force as the only body force, $\vec{g}$ = body force per unit mass, the net force in the x direction can thus be expressed as

$$dF_x = dF_{B_x} + dF_{S_x} = \left(\rho g_x + \frac{\partial \sigma_{xx}}{\partial x} + \frac{\partial \tau_{yx}}{\partial y} + \frac{\partial \tau_{zx}}{\partial z} \right) dx\, dy\, dz \qquad (2.33)$$

We can apply the same procedure to derive the corresponding expressions for the y and z components. The resulting expressions for the net forces in the y and z directions, respectively, are

$$dF_y = dF_{B_y} + dF_{S_y} = \left(\rho g_y + \frac{\partial \tau_{xy}}{\partial x} + \frac{\partial \sigma_{yy}}{\partial y} + \frac{\partial \tau_{zy}}{\partial z} \right) dx\, dy\, dz \qquad (2.34)$$

$$dF_z = dF_{B_z} + dF_{S_z} = \left(\rho g_z + \frac{\partial \tau_{xz}}{\partial x} + \frac{\partial \tau_{yz}}{\partial y} + \frac{\partial \sigma_{zz}}{\partial z} \right) dx\, dy\, dz \qquad (2.35)$$

Substituting Eqs 2.33, 2.34, and 2.35 into Eq. 2.30, noting that $dm = \rho\, dx\, dy\, dz$ and hence the $dx\, dy\, dz$ product on the left-hand side cancels with that on the right-hand side, gives

$$\rho g_x + \frac{\partial \sigma_{xx}}{\partial x} + \frac{\partial \tau_{yx}}{\partial y} + \frac{\partial \tau_{zx}}{\partial z} = \rho \left(\frac{\partial U}{\partial t} + U \frac{\partial U}{\partial x} + V \frac{\partial U}{\partial y} + W \frac{\partial U}{\partial z} \right) \qquad (2.36)$$

$$\rho g_y + \frac{\partial \tau_{xy}}{\partial x} + \frac{\partial \sigma_{yy}}{\partial y} + \frac{\partial \tau_{zy}}{\partial z} = \rho \left(\frac{\partial V}{\partial t} + U \frac{\partial V}{\partial x} + V \frac{\partial V}{\partial y} + W \frac{\partial V}{\partial z} \right) \qquad (2.37)$$

$$\rho g_z + \frac{\partial \tau_{xz}}{\partial x} + \frac{\partial \tau_{yz}}{\partial y} + \frac{\partial \sigma_{zz}}{\partial z} = \rho \left(\frac{\partial W}{\partial t} + U \frac{\partial W}{\partial x} + V \frac{\partial W}{\partial y} + W \frac{\partial W}{\partial z} \right) \qquad (2.38)$$

With F_i as the per unit mass body forces, a continuum fluid with a velocity field U_i, density ρ, and temperature T in the Cartesian coordinate system is

$$\rho \frac{DU_i}{Dt} = \rho F_i + \frac{\partial \sigma_{ji}}{\partial x_j} \qquad (2.39)$$

Alternatively, we can follow Davidson (2004) and apply Newton's second law to a lump of fluid of volume δV to get

$$(\rho \delta V) \frac{D\vec{U}}{Dt} = \oint_S (-P) d\vec{S} + \text{viscous forces} \qquad (2.40)$$

According to Gauss's theorem, we can write

$$\oint_S (-P) d\vec{S} = \int_{\delta V} (-\nabla P) dV = -(\nabla P) \delta V \qquad (2.41)$$

and thus, we have

$$(\rho \delta V) \frac{D\vec{U}}{Dt} = -(\nabla P) \delta V + \text{viscous forces} \qquad (2.42)$$

This equation states that the mass of the fluid element, $\rho \delta V$, times the acceleration, DU/Dt, is equal to the net pressure force acting on the fluid element, plus any viscous forces arising from viscous stresses.

For the infinitesimal fluid lump shown in Fig. 2.2, there are shear and normal stresses as depicted in the figure. Any imbalance in stress will lead to a net force acting on the fluid element. For example, the net force in the x direction is

$$F_x = \left[\frac{\partial \sigma_{xx}}{\partial x} + \frac{\partial \tau_{yx}}{\partial y} + \frac{\partial \tau_{zx}}{\partial z} \right] \delta V \qquad (2.43)$$

This can also be expressed as

$$F_x = \frac{\partial \tau_{jx}}{\partial x_j} \delta V \qquad (2.44)$$

where there is a summation over the repeated index j. Similar expressions can be found for F_y and F_z. The general expression is

$$F_i = \frac{\partial \tau_{ji}}{\partial x_j} \delta V \qquad (2.45)$$

Thus, we can write Eq. 2.42 as

$$\rho \frac{D\vec{U}}{Dt} = -\nabla P + \frac{\partial \tau_{ji}}{\partial x_j} \qquad (2.46)$$

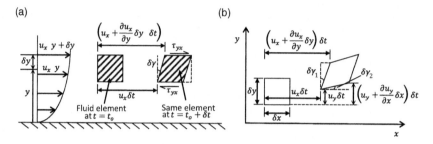

Figure 2.3 Distortion of a fluid element (a) in a parallel shear flow, (b) in a two-dimensional shear flow. *(Created by P.K. Pradip).*

Next, we need a constitutive law relating τ_{ij} to the rate of the deformation of the fluid element. For the fluid element in a parallel shear flow as depicted in Fig. 2.3a, the angular distortion rate, $d\gamma/dt$, is proportional to the shear stress, τ_{yx}, for a Newtonian fluid

$$d\gamma/dt \propto \tau_{yx} \tag{2.47}$$

Introducing the absolute viscosity, $\mu = \rho v$, as the proportionality constant, we have

$$\tau_{yx} = \rho v \, d\gamma/dt \tag{2.48}$$

But $d\gamma/dt = \partial U/\partial y$ and hence, we have

$$\tau_{yx} = \rho v \frac{d\vec{U}}{dy} \tag{2.49}$$

Now we consider the more general two-dimensional case as shown in Fig. 2.3b. Over a small time period δt, the fluid element experiences an angular distortion of

$$\delta\gamma = \delta\gamma_1 + \delta\gamma_2 = (\partial U/\partial y + \partial V/\partial x)\,\delta t \tag{2.50}$$

It follows from Eq. 2.49 that

$$\tau_{xy} = \tau_{yx} = \rho v(\partial U/\partial y + \partial V/\partial x) \tag{2.51}$$

We can easily generalize this into the three-dimensional case; that is, for Newtonian fluids

$$\tau_{ij} = \mu\left(\frac{\partial U_i}{\partial x_j} + \frac{\partial U_j}{\partial x_i}\right) = \rho v\left(\frac{\partial U_i}{\partial x_j} + \frac{\partial U_j}{\partial x_i}\right) \tag{2.52}$$

Substituting this into the equation of motion, Eq. 2.46, gives us the Navier-Stokes equation

$$\frac{D\vec{U}}{Dt} = -\nabla\left(\frac{p}{\rho}\right) + \nu\nabla^2\vec{U} \qquad (2.53)$$

From the material derivative equation, we can express the acceleration of a fluid element as

$$\frac{D\vec{U}}{Dt} = \frac{\partial\vec{U}}{\partial t} + \left(\vec{U}\cdot\nabla\right)\vec{U} \qquad (2.54)$$

With this, we can rewrite the Navier-Stokes equation as

$$\frac{D\vec{U}}{Dt} = -\nabla\left(\frac{p}{\rho}\right) + \nu\nabla^2\vec{U} = \frac{\partial\vec{U}}{dt} + \left(\vec{U}\cdot\nabla\right)\vec{U} \qquad (2.55)$$

Note that for a steady flow, the first term on the right-hand side is zero. On the other hand, the second term is typically finite, for the velocity of the fluid element generally changes as it moves through the flow field.

For incompressible flow with constant viscosity, the corresponding Navier-Stokes equations in the Cartesian coordinate system are

$$\rho g_x - \frac{\partial P}{\partial x} + \mu\left(\frac{\partial^2 U}{\partial x^2} + \frac{\partial^2 U}{\partial y^2} + \frac{\partial^2 U}{\partial z^2}\right) = \rho\left(\frac{\partial U}{\partial t} + U\frac{\partial U}{\partial x} + V\frac{\partial U}{\partial y} + W\frac{\partial U}{\partial z}\right)$$
$$(2.56)$$

$$\rho g_y - \frac{\partial P}{\partial y} + \mu\left(\frac{\partial^2 V}{\partial x^2} + \frac{\partial^2 V}{\partial y^2} + \frac{\partial^2 V}{\partial z^2}\right) = \rho\left(\frac{\partial V}{\partial t} + U\frac{\partial V}{\partial x} + V\frac{\partial V}{\partial y} + W\frac{\partial V}{\partial z}\right)$$
$$(2.57)$$

$$\rho g_z - \frac{\partial P}{\partial z} + \mu\left(\frac{\partial^2 W}{\partial x^2} + \frac{\partial^2 W}{\partial y^2} + \frac{\partial^2 W}{\partial z^2}\right) = \rho\left(\frac{\partial W}{\partial t} + U\frac{\partial W}{\partial x} + V\frac{\partial W}{\partial y} + W\frac{\partial W}{\partial z}\right)$$
$$(2.58)$$

2.4 REYNOLDS DECOMPOSITION

In this section, we are going to follow Reynolds (1895) in the decomposition of the Navier-Stokes equation with the Reynolds-averaged Navier-Stokes equation. Strictly speaking, we are invoking the statistical theory for continuum turbulent flow when we talk about Reynolds averaging. In other words, since it is impossible for us to develop a deterministic theory of turbulence, we opt to develop a statistical one based on the average properties of turbulence. As soon as we talk about averages, we have to know what kind of average we are dealing with (Choudhuri, 1998).

For spatially homogeneous turbulence, we can take spatial averages of various fluctuating quantities over some region of space. For "stationary" turbulence in which its general (statistical) characteristics are invariant with respect to time, we can take time averages. Perhaps ensemble averages are the most general kind of average. One can think of the cyclic nature of the in-cylinder turbulent flow in a reciprocating engine. The repeating cycles are replicas of the same system having the same statistical properties of turbulence, though the actual value of quantities like velocity at the same spatial point at the same crank angle in the different members (cycles) of the ensemble may be different. By averaging the values of the same quantity in different members of the ensemble, we can obtain the ensemble average.

Then there is the vexing question of ergodicity. If there is more than one kind of averaging, are the different averaging procedures equivalent? Let us consider a fluctuating velocity U_i which can be decomposed into a steady (or a slowly varying "mean") component, plus a fluctuating component as sketched in Fig. 2.4, where U_i = instantaneous value of the ith velocity, $\bar{U}_i$ = time-averaged value and, ($u_i = U_i - \bar{U}_i$), fluctuating component.

The general time-averaged velocity is

$$\bar{U}_i(t_i) = \frac{1}{T_{\text{period}}} \int_{t_1 - T_{\text{period}}/2}^{t_1 + T_{\text{period}}/2} U_i(t)\, dt \qquad (2.59)$$

This is applicable for "stationary" and slowly varying "mean" turbulent flows. In other words, the average is meaningful if the variation in the mean velocity is relatively slow and small within an adequately long time period T_{period} over which the average is deduced.

If the mean velocity $\bar{U}_i$ is repeatable, such as that portrayed in Fig. 2.5, we can invoke ensemble averaging, that is

$$\bar{U}_i(t_1) = \langle U_i \rangle \equiv \frac{1}{N} \sum_{n=1}^{N} U_{i,n}(t_1) \qquad (2.60)$$

Here, N is the total number of cycles, n is the cycle number, and subscript "i" signifies the crank angle degree, for example.

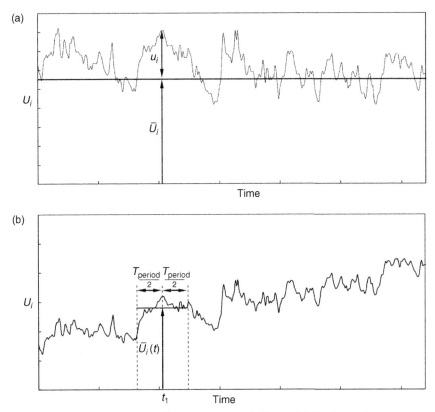

Figure 2.4 Velocity time trace of (a) stationary turbulence, (b) slowly varying mean turbulence. *(Created by A.R. Vasel-Be-Hagh).*

If the ensemble mean $<U>$ is independent of time, the process is stationary. For stationary processes, it can be shown that

$$\overline{U} = \langle U \rangle \tag{2.61}$$

The time mean or average is

$$\overline{U} \equiv \lim_{T_{period} \to \infty} \frac{1}{T_{period}} \int_0^{T_{period}} U(t)\, dt \tag{2.62}$$

Alternatively, the average can be invoked from minus to plus one half the time period, as per Eq. 2.59.

It is worth mentioning some of the basic rules of time averaging here. An averaged quantity in a stationary process is taken as a constant (with respect to time) in the next average. This can all be proved using an overbar,

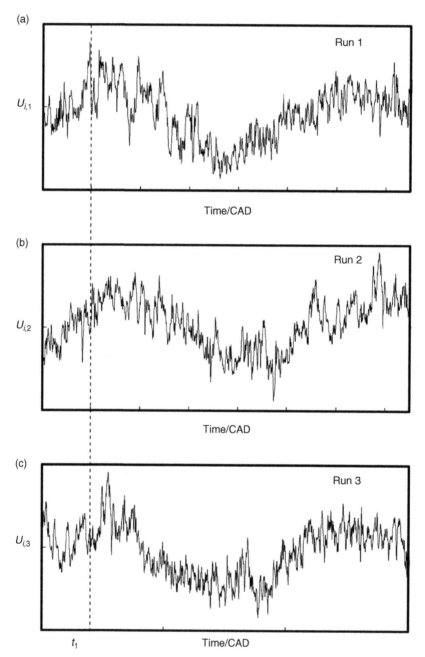

Figure 2.5 Cyclic turbulent flow motion (CAD = crank angle degree). Three cycles, (a) Run 1, (b) Run 2, (c) Run 3 are shown. *(Created by A.R. Vasel-Be-Hagh).*

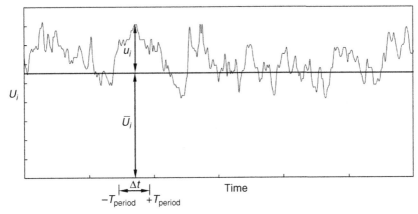

Figure 2.6 A "stationary" turbulent flow. *(Created by A.R. Vasel-Be-Hagh).*

$" ^- " = \dfrac{1}{T_{period}} \displaystyle\int_0^{T_{period}} dt$. Let us look at a stationary velocity fluctuation, such as that shown in Fig. 2.6, where

$$U = \bar{U} + u \qquad (2.63)$$

Taking the average, we get

$$\overline{U} = \overline{\bar{U} + u} = \overline{\bar{U}} + \bar{u} \qquad (2.64)$$

But $\bar{U} = \overline{\bar{U}}$, and hence, $\bar{u} = 0$.

2.5 CONSERVATION OF MASS FROM LAMINAR TO TURBULENT FLOW

From the general mass conservation expression such as Eq. 2.27, we have

$$\frac{\partial \rho}{\partial t} + \frac{\partial \left(\rho U_j \right)}{\partial x_j} = 0 \qquad (2.65)$$

Expanding the terms in the bracket, we get

$$\frac{\partial \rho}{\partial t} + U_j \frac{\partial \rho}{\partial x_j} + \rho \frac{\partial U_j}{\partial x_j} = 0 \qquad (2.66)$$

For (steady) laminar flows, the instantaneous velocity is equal to the mean or time-averaged velocity, that is, $U_j = \bar{U}_j$. In other words, the continuity is as simple as is expressed in the equations mentioned earlier.

For turbulent flows, on the other hand, the instantaneous velocity consists of a fluctuating component in addition to a mean velocity. The mean velocity is a constant for "stationary" or "steady" flows. Let us decompose the (total) instantaneous velocity into the mean and a randomly fluctuating component, that is

$$U_j = \bar{U}_j + u_j \tag{2.67}$$

To illustrate a couple of points concerning averaging of the product of two randomly varying parameters with otherwise steady averaged values, let us consider the density of the fluid to be composed of a mean and a randomly fluctuating value, that is

$$\rho = \bar{\rho} + \tilde{\rho} \tag{2.68}$$

Substitute Eqs 2.67 and 2.68 into Eq. 2.65, and we have

$$\frac{\partial(\bar{\rho} + \tilde{\rho})}{\partial t} + \frac{\partial\left[(\bar{\rho} + \tilde{\rho})(\bar{U}_j + u_j)\right]}{\partial x_j} = 0. \tag{2.69}$$

Take the average

$$\frac{\partial(\overline{\bar{\rho} + \tilde{\rho}})}{\partial t} + \frac{\partial\left[\overline{(\bar{\rho} + \tilde{\rho})(\bar{U}_j + u_j)}\right]}{\partial x_j} = 0 \tag{2.70}$$

The averages of the terms consisting of only one randomly fluctuating parameter are all zeros, leaving us with

$$\frac{\partial\bar{\rho}}{\partial t} + \frac{\partial\left(\bar{\rho}\bar{U}_j + 0 + 0 + \overline{\tilde{\rho}u_j}\right)}{\partial x_j} = 0 \tag{2.71}$$

The terms with mean parameters, indicated with an overbar, are the same as those for the laminar flow. We see an additional term (compared to the laminar case), which indicates a possible "correlation" between the density fluctuation and the velocity fluctuation. A simple example of this averaging of a product of two time-varying parameters is depicted in Fig. 2.7. It is clear that $\overline{\tilde{\rho}u_j}$ is not necessarily zero, and thus, cannot be assumed so.

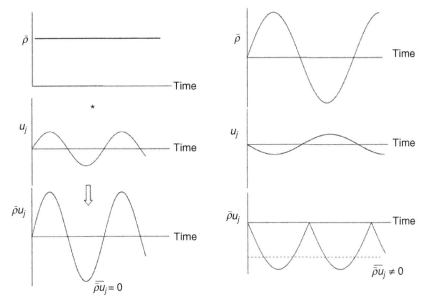

Figure 2.7 Averaging of the product of time-varying parameters. *(Created by M. Ahmadi-Baloutaki).*

2.6 MOMENTUM EQUATION IN TURBULENT FLOW

In this section, we follow Reynolds decomposition and decompose the involved parameters into a steady (mean) and a randomly fluctuating part and take the average (Wilson, 1989). While the resulting final expressions are readily available in the open literatures, especially in the authoritative "bible" by Hinze (1959, 1975) and in monographs such as Wilcox (2006) and Garde (2010), we will walk through the derivation step by step to clearly show where the additional terms, as compared to the laminar case, come from.

Consider the x-direction momentum equation for the instantaneous velocity U for a constant density and constant viscosity flow, that is

$$\frac{\partial U}{\partial t} + U\frac{\partial U}{\partial x} + V\frac{\partial U}{\partial y} + W\frac{\partial U}{\partial z} = -\frac{1}{\rho}\frac{\partial P}{\partial x} + v\left(\frac{\partial^2 U}{\partial x^2} + \frac{\partial^2 U}{\partial y^2} + \frac{\partial^2 U}{\partial z^2}\right) \quad (2.72)$$

This can be expressed, alternatively as

$$\frac{\partial U_i}{\partial t} + U_j\frac{\partial U_i}{\partial x_j} = -\frac{1}{\rho}\frac{\partial P}{\partial x_i} + v\frac{\partial^2 U_i}{\partial x_j \partial x_j} \quad (2.73)$$

Executing Reynolds decomposition, we have

$$U = \bar{U} + u \tag{2.74}$$

$$V = \bar{V} + v \tag{2.75}$$

$$W = \bar{W} + w \tag{2.76}$$

$$P = \bar{P} + p \tag{2.77}$$

Here the small letters are used to signify the randomly fluctuating components. For the x direction, that is, $i = 1$, we have

$$\frac{\partial(\bar{U} + u)}{\partial t} + (\bar{U} + u)\frac{\partial(\bar{U} + u)}{\partial x} + (\bar{V} + v)\frac{\partial(\bar{U} + u)}{\partial y} + (\bar{W} + w)\frac{\partial(\bar{U} + u)}{\partial z}$$
$$= -\frac{1}{\rho}\frac{\partial(\bar{P} + p)}{\partial x} + v\left[\frac{\partial^2(\bar{U} + u)}{\partial x^2} + \frac{\partial^2(\bar{U} + u)}{\partial y^2} + \frac{\partial^2(\bar{U} + u)}{\partial z^2}\right] \tag{2.78}$$

Considering the steady flow case, where $\dfrac{\partial U}{\partial t} = \dfrac{\partial(\bar{U} + u)}{\partial t} = 0$, we can take the average of the aforementioned equation, that is

$$\overline{(\bar{U} + u)\frac{\partial(\bar{U} + u)}{\partial x}} + \overline{(\bar{V} + v)\frac{\partial(\bar{U} + u)}{\partial y}}$$
$$+ \overline{(\bar{W} + w)\frac{\partial(\bar{U} + u)}{\partial z}} = -\frac{1}{\rho}\frac{\partial\bar{P}}{\partial x} + v\left(\frac{\partial^2\bar{U}}{\partial x^2} + \frac{\partial^2\bar{U}}{\partial y^2} + \frac{\partial^2\bar{U}}{\partial z^2}\right) \tag{2.79}$$

Here, we have further simplified the considered case by assuming that there is no pressure fluctuation for this constant density (incompressible) flow.

The nonlinear acceleration terms in Eq. 2.79 can be expanded as follows

$$\overline{(\bar{U} + u)\frac{\partial(\bar{U} + u)}{\partial x}} = \bar{U}\frac{\partial\bar{U}}{\partial x} + \overline{\bar{U}\frac{\partial u}{\partial x}} + \overline{u\frac{\partial\bar{U}}{\partial x}} + \overline{u\frac{\partial u}{\partial x}} = \bar{U}\frac{\partial\bar{U}}{\partial x} + \overline{u\frac{\partial u}{\partial x}} \tag{2.80}$$

$$\overline{(\bar{V} + v)\frac{\partial(\bar{U} + u)}{\partial y}} = \bar{V}\frac{\partial\bar{U}}{\partial y} + \overline{\bar{V}\frac{\partial u}{\partial y}} + \overline{v\frac{\partial\bar{U}}{\partial y}} + \overline{v\frac{\partial u}{\partial y}} = \bar{V}\frac{\partial\bar{U}}{\partial y} + \overline{v\frac{\partial u}{\partial y}} \tag{2.81}$$

$$\overline{\left(\bar{W}+w\right)\frac{\partial\left(\bar{U}+u\right)}{\partial z}} = \bar{W}\frac{\partial\bar{U}}{\partial z}+\bar{W}\frac{\partial\bar{u}}{\partial z}+\bar{w}\frac{\partial\bar{U}}{\partial z}+\overline{w\frac{\partial u}{\partial z}} = \bar{W}\frac{\partial\bar{U}}{\partial z}+\overline{w\frac{\partial u}{\partial z}} \qquad (2.82)$$

With these expanded terms we can rewrite Eq. 2.79 as

$$\bar{U}\frac{\partial\bar{U}}{\partial x}+\overline{u\frac{\partial u}{\partial x}}+\bar{V}\frac{\partial\bar{U}}{\partial y}+\overline{v\frac{\partial u}{\partial y}}+\bar{W}\frac{\partial\bar{U}}{\partial z}+\overline{w\frac{\partial u}{\partial z}}$$
$$= -\frac{1}{\rho}\frac{\partial\bar{P}}{\partial x}+\nu\left(\frac{\partial^2\bar{U}}{\partial x^2}+\frac{\partial^2\bar{U}}{\partial y^2}+\frac{\partial^2\bar{U}}{\partial z^2}\right) \qquad (2.83)$$

which can be rearranged into

$$\bar{U}\frac{\partial\bar{U}}{\partial x}+\bar{V}\frac{\partial\bar{U}}{\partial y}+\bar{W}\frac{\partial\bar{U}}{\partial z}+\left(\overline{u\frac{\partial u}{\partial x}}+\overline{v\frac{\partial u}{\partial y}}+\overline{w\frac{\partial u}{\partial z}}\right)$$
$$= -\frac{1}{\rho}\frac{\partial\bar{P}}{\partial x}+\nu\left(\frac{\partial^2\bar{U}}{\partial x^2}+\frac{\partial^2\bar{U}}{\partial y^2}+\frac{\partial^2\bar{U}}{\partial z^2}\right) \qquad (2.84)$$

Recall that the continuity equation for an incompressible flow can be expressed as

$$\frac{\partial U}{\partial x}+\frac{\partial V}{\partial y}+\frac{\partial W}{\partial z}=0 \qquad (2.85)$$

where U, V, and W signify the instantaneous velocities in the three orthogonal directions of the Cartesian coordinates. Invoking Reynolds decomposition followed by averaging

$$\frac{\partial\overline{\left(\bar{U}+u\right)}}{\partial x}+\frac{\partial\overline{\left(\bar{V}+v\right)}}{\partial y}+\frac{\partial\overline{\left(\bar{W}+w\right)}}{\partial z}=0 \qquad (2.86)$$

we get

$$\frac{\partial\bar{U}}{\partial x}+\frac{\partial\bar{V}}{\partial y}+\frac{\partial\bar{W}}{\partial z}=0 \qquad (2.87)$$

Subtract this from the instantaneous continuity equation, that is, Eq. 2.85 minus Eq. 2.87, to obtain the turbulence

$$\frac{\partial u}{\partial x}+\frac{\partial v}{\partial y}+\frac{\partial w}{\partial z}=0 \qquad (2.88)$$

Then, multiply this by the fluctuating velocity in the x direction, u, and take (time) average to obtain

$$\overline{u\frac{\partial u}{\partial x}} + \overline{u\frac{\partial v}{\partial y}} + \overline{u\frac{\partial w}{\partial z}} = 0 \tag{2.89}$$

Add this zero to the left-hand side of the momentum equation, Eq. 2.84, and note that

$$\overline{u\frac{\partial v}{\partial y}} + \overline{v\frac{\partial u}{\partial y}} = \frac{\partial \overline{uv}}{\partial y} \tag{2.90}$$

The left-hand side of the momentum equation becomes

$$\bar{U}\frac{\partial \bar{U}}{\partial x} + \bar{V}\frac{\partial \bar{U}}{\partial y} + \bar{W}\frac{\partial \bar{U}}{\partial z} + \overline{u\frac{\partial u}{\partial x}} + \overline{v\frac{\partial u}{\partial y}} + \overline{w\frac{\partial u}{\partial z}} + \overline{u\frac{\partial u}{\partial x}} + \overline{u\frac{\partial v}{\partial y}} + \overline{u\frac{\partial w}{\partial z}}$$

$$= \bar{U}\frac{\partial \bar{U}}{\partial x} + \bar{V}\frac{\partial \bar{U}}{\partial y} + \bar{W}\frac{\partial \bar{U}}{\partial z} + \frac{\partial \overline{u^2}}{\partial x} + \frac{\partial \overline{uv}}{\partial y} + \frac{\partial \overline{uw}}{\partial z} \tag{2.91}$$

Transposing these terms in the momentum equation yields

$$\rho\left(\bar{U}\frac{\partial \bar{U}}{\partial x} + \bar{V}\frac{\partial \bar{U}}{\partial y} + \bar{W}\frac{\partial \bar{U}}{\partial z}\right) = -\frac{\partial P}{\partial x} + \frac{\partial}{\partial x}\left(\mu\frac{\partial \bar{U}}{\partial x} - \rho\overline{u^2}\right)$$

$$+ \frac{\partial}{\partial y}\left(\mu\frac{\partial \bar{U}}{\partial y} - \rho\overline{uv}\right) + \frac{\partial}{\partial z}\left(\mu\frac{\partial \bar{U}}{\partial z} - \rho\overline{uw}\right) \tag{2.92}$$

Comparing this to the corresponding laminar counterpart, we see three additional terms involving the fluctuating velocities u, v, and w. These extra stresses, as illustrated in Fig. 2.8, are interpreted as "Reynolds stresses."

To help us to comprehend Reynolds stresses, let us look at a one-dimensional shear flow as portrayed in Fig. 2.9, where the lower, slower-moving fluid drags the upper, faster-moving fluid. If we define viscous shear as

$$\tau_{vis} \equiv \mu\frac{d\vec{U}}{dy} > 0 \tag{2.93}$$

we see that $\tau_{vis} > 0$ (positive shear) points to the left.

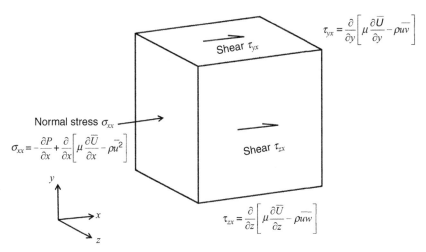

Figure 2.8 Reynolds stresses. *(Created by H. Can).*

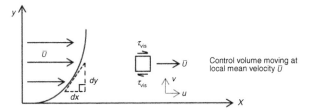

Figure 2.9 One-dimensional shear flow in a boundary layer. *(Created by M. Ahmadi-Baloutaki).*

Further, let us define τ'_{turb} as an apparent fluctuating turbulent "stress" that causes the same effect as the momentum added by turbulence. This may be somewhat illustrated by Zorro dashing by on his swift horse and jumping onto a slower, steadily moving wagon; see Fig. 2.10. The steadily moving wagon signifies a fluid particle moving at the mean velocity, where the momentum is unchanging with respect to time. On the other hand, the dashing Zorro portrays turbulent fluctuation (momentum) of the otherwise steadily moving fluid particle.

Following the analogy depicted in Fig. 2.10, the force-momentum balance can be expressed as

$$-\tau'_{turb}\, dA \equiv (\rho v dA)u \tag{2.94}$$

where A is area, and u and v are the fluctuating velocities in the x and y directions, respectively. The negative sign evolves from the fact that $a + u$

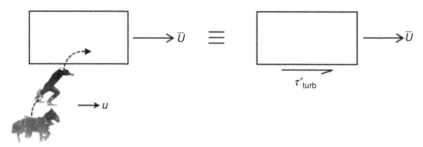

Figure 2.10 A dashing Zorro jumping onto a slower moving wagon as an illustration of fluctuating turbulent stress. *(Created by M. Ahmadi-Baloutaki).*

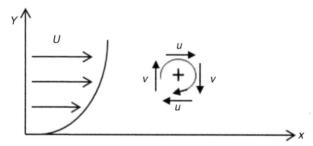

Figure 2.11 A two-dimensional correlation of fluctuating turbulent velocities in boundary shear. *(Created by M. Ahmadi-Baloutaki).*

causes a τ'_{turb} in the $+x$ direction, which is negative τ. The above equation can be simplified into

$$\tau'_{turb} = -\rho uv \tag{2.95}$$

Taking a time average, where $\overline{\tau'_{turb}} \equiv \tau_{turb}$, we have

$$\tau_{turb} = -\rho\overline{uv} \tag{2.96}$$

Note that the generation of apparent shear stress by turbulence requires u to be "correlated" with v such that $\overline{uv} \neq 0$. Figure 2.11 shows that in boundary layer flows when $u < 0$, we tend to have $v > 0$ and when $u > 0$, we tend to see that $v < 0$; therefore, $\overline{uv} < 0$ and $\tau_{turb} > 0$.

In a jet flow, on the other hand, $\overline{uv} > 0$ so $\tau_{turb} < 0$; see Fig. 2.12. But in a jet flow, we also see that $\partial U/\partial y < 0$, which suggests there may be an effective turbulent "eddy viscosity," μ_{turb}, that is

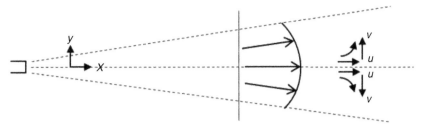

Figure 2.12 A two-dimensional jet showing the correlation of the fluctuating turbulent velocities. *(Created by M. Ahmadi-Baloutaki).*

$$\tau_{turb} = -\rho\overline{uv} \equiv \mu_{turb}\frac{\partial U}{\partial y} \tag{2.97}$$

We can rewrite this as

$$-\overline{uv} \equiv \nu_{turb}\frac{\partial U}{\partial y} \tag{2.98}$$

where $\nu_{turb}(x_i, U_i, u_i)$ is a turbulent momentum exchange coefficient which is *not constant* in x, y, z like $\nu = \mu/\rho$ is.

In short, the above simplified one-dimensional shear flow illustrations show that there is an additional stress caused by the fluctuating component of the flow. The final equation for the total shear stress on the x-y surface of the control volume depicted in Fig. 2.8 is

$$\tau_{xy} = \mu\frac{\partial U}{\partial y} - \rho\overline{uv} \tag{2.99}$$

Problems

Problem 2.1. Energy equation for an ideal gas
Derive the energy equation for an ideal, incompressible gas. Use standard notations and show all steps clearly.

Problem 2.2. Applicable equations and parameters in two-dimensional flows
What are the unknowns and the required equations for a two-dimensional, unsteady a) incompressible flow, and b) compressible flow?

Problem 2.3. Applicable equations and parameters in three-dimensional flows
What are the unknowns and the required equations for a three-dimensional, unsteady, a) incompressible flow, and b) compressible flow?

Problem 2.4. Navier-Stokes equations for compressible turbulent flows of an ideal gas

Derive the Navier-Stokes equations for compressible turbulent flows of an ideal gas.

Problem 2.5. Reynolds stresses

A blob of dye is dropped into a pool of agitated water. Assume the blob is a sphere in the middle of the pool of water and that the pool of water has a zero mean velocity. How do the Reynolds stresses affect the dispersion of the blob of dye?

Problem 2.6. Sampling oscillating flow

The velocity of moving water in a water channel is controlled such that its magnitude oscillates between 0.5 m/s and 1 m/s periodically at 0.25 Hz. A hot-film is used to quantify the flow. What should the sample rate and sample size be? How can you verify that these settings are sufficient? Back up your solution using plots, etc.

REFERENCES

Çengel, Y.A., Cimbala, J.M., 2013. Fluid Mechanics: Fundamentals and Applications, third ed. McGraw-Hill, New York.

Choudhuri, A.R., 1998. The Physics of Fluids and Plasmas. Cambridge University Press, Cambridge.

Currie, I.G., 1974. Fundamental Mechanics of Fluids. McGraw-Hill, New York.

Davidson, P.A., 2004. Turbulence: An Introduction for Scientists and Engineers. Oxford University Press, New York.

Fox, R.W., Pritchard, P.J., McDonald, A.T., 2009. Introduction to Fluid Mechanics, seventh ed. John Wiley & Sons, USA.

Garde, R.J., 2010. Turbulent Flow, third ed. New Age Science, UK.

Haller, G., 2015. Lagrangian coherent structures. J. Fluid Mech. 47, 137–162.

Hinze, J.O., 1959. Turbulence. McGraw-Hill, USA.

Hinze, J.O., 1975. Turbulence, second ed. McGraw-Hill, USA.

Reynolds, O., 1895. On the dynamical theory of incompressible viscous fluids and the determination of the criterion. Philos. Trans. R. Soc. Lond. Ser. A 186, 123–164.

Wilcox, D.C., 2006. Turbulence Modeling for CFD, third ed. DCW, USA.

Wilson, D.J., 1989. Mec E 632: Turbulent Fluid Dynamics, Lecture Notes, University of Alberta, Edmonton.

CHAPTER 3

Statistical Description of Flow Turbulence

You can never cross the ocean unless you have the courage to lose sight of the shore.

– Christopher Columbus

Contents

Chapter Objectives

- To review basic statistical terms and analyses.
- To understand first, second, third, and fourth central moments.
- To appreciate flow turbulence based on these statistical premises.
- To comprehend correlations and covariances of the turbulent velocity.
- To deduce integral and Taylor micro scales from autocorrelation.

NOMENCLATURE

A	An event
C	A condition
D	Diameter
F	Cumulative distribution function
f	Probability density function
K	Kurtosis or flatness factor, $K \equiv \overline{u^4}/\sigma^4$
OPP	Orificed, Perforated, Plate
p, P	Probability
PDF	Probability density function
r	Spatial distance

Basics of Engineering Turbulence
http://dx.doi.org/10.1016/B978-0-12-803970-0.00003-9

R Covariance
rms   Root mean square
S Skewness factor, $S \equiv \overline{u^3}/\sigma^3$
t Time
U Instantaneous velocity in the x (streamwise) direction
u The fluctuating velocity component in the x (streamwise) direction
v The fluctuating velocity component in the y (transverse) direction
x, y, z Cartesian coordinates
Λ Integral scale
λ (Taylor) microscale
ρ Autocorrelation
τ Time, time scale, time interval
σ Root mean square, $\sigma = \langle u^2 \rangle = \sqrt{\overline{u^2}}$, standard deviation
$< >$ Average

3.1 INTRODUCTION

On one hand, the detailed behaviors of random flow turbulence are un-predictable; on the other hand, several statistical characteristics of the flow are largely reproducible. In other words, the random character of turbulent flows strongly suggests that statistical methods will be useful. Taylor realized this in the early 1900s and contributed significantly to viewing turbulence from a statistical perspective (Taylor, 1935, 1936). There are, however, also those who are less optimistic about the statistical approach for studying turbulence. They suggest there is a limit which statistical approach cannot surpass. Even if this is the case, looking at flow turbulence through the statistical window is surely beneficial in comprehending the mystifying phenomenon of flow. We will start this chapter with a brief review of the basic statistics used for describing a random variable. The fluctuating turbulent velocity is the random variable of concern, and its particular behaviors as described by the various statistical factors covered will be explained.

A random process is "stationary" when its statistical characteristics are not changing, though its instantaneous value varies randomly with respect to time. To stay in context, consider the instantaneous velocity of the stationary turbulent flow, shown in Fig. 3.1, as the random variable. That an event is called "random" implies that it is neither certain nor impossible. It may occur but need not occur. For illustration purposes, we assume a random event, Event A, to be the situation when the instantaneous velocity U is between 0.90 and 0.93 times the average (mean) velocity $\overline{U}$. The instantaneous velocity

$$U = \overline{U} + u \tag{3.1}$$

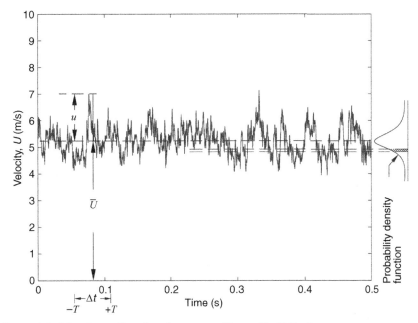

Figure 3.1 A "stationary" random flow event. *(Created by N. Cao).*

where $\bar{U}$ = mean velocity, and u = fluctuating velocity. Velocity U is a random variable, as it does not have a unique value, that is, the same value every time the experiment is repeated under the same set of conditions, C. For the considered case, the instantaneous velocity does not have a unique value at any given time instant, t. In other words, there are always perturbations in reality; hence, we can never repeat the same exact initial and/or boundary conditions C in any two realizations. Coming back to Event A, the probability for it to occur is the sum of the length of time that the instantaneous velocity is between $0.90\ \bar{U}$ and $0.93\ \bar{U}$, divided by the total time period considered. This is depicted by the hatched slice on the bell-shaped probability density curve, which will be expounded shortly. The hatched area divided by the total area under the probability density curve is the probability for Event A to occur.

It is interesting to note that even though the equations of motion, the Navier-Stokes equations, which were introduced in the previous chapter, are deterministic, the solutions are random. The randomness is a result of the unavoidable perturbations in the initial conditions, boundary conditions, and material properties. Turbulent flow fields display an acute sensitivity to such perturbations. In fact, at high Reynolds numbers, the flow is particularly sensitive to these small perturbations. This can be illustrated using, for example, the Lorenz equations (Moon, 1992, among others). Let us recap the "butterfly effect" according to chaos theory. The faint flapping of a butterfly near a tree next to a river in the Amazon rainforest can lead to the falling of a leaf. This leaf, which falls into the river can marginally alter the flow stream. The alteration of the flow stream can result in some change in the rain formation. As the effect propagates, it also escalates and eventually could manifest itself into a hurricane off the east coast of Canada. Turbulent flows are thus a very sensitive type of flow motion.

By looking at the statistical properties of an ensemble of different flow realizations, all obtained using the same nominal conditions, one hopes to extract useful quantities such as probabilities and averages, which depend only on parameters controlled by the experimenter. For example, the average velocity as depicted in Fig. 3.1 is well-defined, despite significant instantaneous fluctuations. The departure of any given realization from the mean can be calculated by subtracting the mean value from the total; this is conventionally identified as turbulence.

When studying turbulence, it is important to know a few things. First, we must know how fluctuations are distributed around an average value. This requires the use of probability density and its Fourier transform, as well as the characteristic function. Secondly, we also need the central limit theorem for making the shape of the probability density of certain quantities. Thirdly, we must know how adjacent fluctuations next to each other in time and/or space are related. For the last point, we require the autocorrelation and its Fourier transform, also known as the energy spectrum. Before we elaborate any further, let us refresh our understanding of probability.

3.2 PROBABILITY

The probability of a random event, Event A for example, can be written as

$$p = P(A) = P\{4.8\,\text{m/s} \le U < 4.9\,\text{m/s}\} \qquad (3.2)$$

where p is a real number between 0 and 1, that is, $0 \le p \le 1$. For an impossible event such as the universe revolving around the earth, $p = 0$. On the other hand, for a sure event such as the sun rising from the east, $p = 1$.

Consider a random signal as shown in Fig. 3.2a, where the signal fluctuates between −2 and 2 over a three-second time period. The *cumulative distribution function* denotes the probability that the random signal will be less

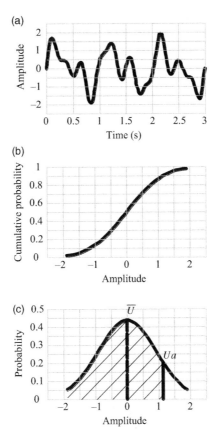

Figure 3.2 A random time series, cumulative distribution, and probability density function. (a) Turbulent wave, (b) cumulative density function, (c) probability density function. *(Created by H. Cen).*

than a particular value. For example, the cumulative distribution function for velocity U to be less than U_a is

$$F(U_a) \equiv P\{U < U_a\}, \text{ or, } P\{-\infty \leq U < U_a\} \tag{3.3}$$

where U_a is the numerical value of the upper bound; see Fig. 3.2b. We also recognize that the probability of variable U having a value between U_a and U_b can be deduced from the cumulative distribution functions

$$F(U_b) - F(U_a) = P\{U_a \leq U < U_b\} \tag{3.4}$$

Since it is impossible for the variable to be less than negative infinity, $U < -\infty$, we have

$$F(-\infty) = 0 \tag{3.5}$$

When the cutoff value for U is infinity, that is for $U < \infty$, we have included all possible values that U could have; hence

$$F(\infty) = 1 \tag{3.6}$$

According to Fig. 3.2b, which is a plot of the *cumulative density function*, the probability that the signal is less than or equal to 0.5 is approximately 0.7; in other words, there is a 70% chance that the amplitude of the fluctuation is no more than 0.5. This 70% probability is also depicted by the hatched area under the *probability density function* (PDF) in Fig. 3.2c.

We observed from Fig. 3.2 that the PDF is defined as the derivative of the cumulative density function

$$f(U) = \frac{dF(U)}{dU} \tag{3.7}$$

We note that since the PDF must be non-negative, we have

$$f(U) \geq 0 \tag{3.8}$$

When all possibilities are included, the probability is 100%, that is

$$\int_{-\infty}^{\infty} f(U)\,dU = 1 \tag{3.9}$$

The probability of a random variable having a value between an interval is equal to the integral of the PDF over that interval. For example, the probability for U to have a value between U_a and U_b is

$$P\{U_a \leq U < U_b\} = F(U_b) - F(U_a) = \int_{U_a}^{U_b} f(U)\,dU \tag{3.10}$$

Hence, we learn that the PDF, $f(U)$, is the probability per unit distance in the sample space. This implies that

$$f(U) = P\{\}/dU \tag{3.11}$$

Let us restrict the discussion to "stationary" turbulent flows, where the fluctuating quantities are statistically steady, as illustrated in Fig. 3.3. The relative amount of time that $U(t)$ spends at various levels is the probability density $f(U)$, that is

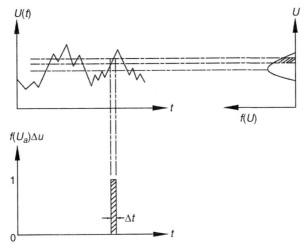

Figure 3.3 Creation of the probability density function. *(Created by F. Iakovidis).*

$$f(U)\Delta U \equiv \lim_{T \to \infty} \frac{1}{T} \sum (\Delta t) \tag{3.12}$$

which is the probability of finding $U(t)$ between U and $U + \Delta U$, or the proportion of time $U(t)$ spent in that interval.

One interesting example for illustrating probability is the amount of time that a pendulum spends at any position between the two extremes. The displacement (velocity) of a pendulum is traced in Fig. 3.4, along with the corresponding PDF. We see that in the case of a pendulum, like many human beings, most of the is squandered at the extremes; that is, most of us spend very little time at the well-balanced equilibrium position of moderation.

For an arbitrary function $g(U)$, the time average is

$$\overline{g} = \lim_{T \to \infty} \frac{1}{T} \int_0^T g(U)\, dt \tag{3.13}$$

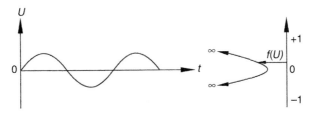

Figure 3.4 Probability density function of a pendulum. *(Created by F. Iakovidis).*

It is worth noting that this time average can be formed by adding all the time intervals between $t = 0$ and $t = T$ over which $U(t)$ is between U and $U + \Delta U$, multiplying this by $g(U)$, and then find their sum, that is

$$\bar{g} = \int_{-\infty}^{\infty} g(U) f(U) \, dU. \tag{3.14}$$

3.3 MOMENTS

In statistics, moments are the mean values of the various powers of the variables of question. Staying in the flow turbulence context, let us follow Batchelor (1953), Tennekes and Lumley (1972), Flierl and Ferrari (2007), and Garde (2010) and consider flow velocity as the random variable. The first moment is simply the mean or time-averaged velocity, and according to Eq. (3.14)

$$\bar{U} \equiv \int_{-\infty}^{\infty} U f(U) \, dU \tag{3.15}$$

where the instantaneous velocity consists of the mean plus the fluctuating velocities, that is, $U = \bar{U} + u$. The PDF of the instantaneous velocity $f(U)$ is equal to $f(\bar{U} + u)$. Thus, we can obtain the PDF of the fluctuating component $f(u)$ simply by shifting the function over a distance $\bar{U}$ along the U-axis so that the new mean is zero. This is similar to subtracting the mean or time-averaged velocity $\bar{U}$ from the total velocity U, which leaves us with only the fluctuating velocity u; we then deduce its probability function $f(u)$. Let us continue to focus on the perplexing fluctuating component u, which is the flow turbulence without the mean convecting velocity.

The moments formed with u^n and $f(u)$ are called *central moments*; see Fig. 3.5. The nth central moment $<u^n>$ is defined as

$$\langle u^n \rangle \equiv \int_{-\infty}^{\infty} u^n f(u) \, du \tag{3.16}$$

We can see from Fig. 3.5 that the *first central moment*

$$< u > = \bar{u} = 0 \tag{3.17}$$

for the mean value of a randomly fluctuating velocity, which is simplified as a simple sine wave in Fig. 3.5, is zero. In other words, when the mean velocity is removed, the mean of the remaining fluctuating velocity is zero.

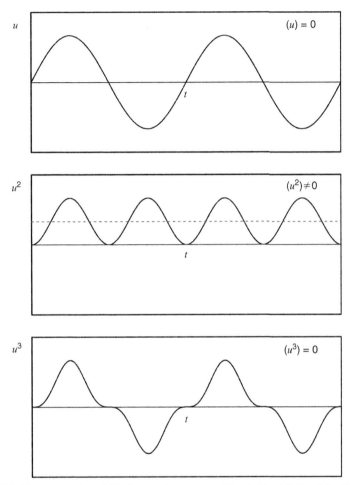

Figure 3.5 First, second, and third central moments. *(Created by A.R. Vasel-Be-Hagh).*

The *second central moment* is the mean-square departure from the mean value $\overline{U}$. It is called the variance and is defined as

$$\overline{u^2} \equiv \int_{-\infty}^{\infty} u^2 f(u)\, du \qquad (3.18)$$

The square root of the variance, $\sigma = \sqrt{\overline{u^2}}$, is the familiar standard deviation, or root-mean-square (rms) amplitude of the fluctuation u_{rms}. This standard deviation is the most convenient measure of the width of probability density function $f(u)$. We clearly see that the standard deviation of the sine wave portrayed in Fig. 3.5 is not zero. In fact, it is only zero when there is no fluctuation, that is, when u is equal to zero. Furthermore, it is worth noting that neither σ nor σ^2 is affected by any lack of symmetry in $f(u)$.

The *third central moment*

$$\overline{u^3} \equiv \int_{-\infty}^{\infty} u^3 f(u)\, du \qquad (3.19)$$

signifies the amount of skewness in the PDF $f(u)$. This measure of the lack of symmetry is commonly expressed in the normalized form called the skewness factor

$$S \equiv \overline{u^3} / \sigma^3 \qquad (3.20)$$

The smaller the value of S, the more symmetrical the probability distribution of u. A perfect, symmetrically distributed u such as the sine wave in Fig. 3.5 gives $S = \langle u^3 \rangle = 0$. On the other hand, Fig. 3.6 portrays a positively skewed signal. The positively skewed signal in Fig. 3.6 indicates that the variable fluctuates much farther, though sparsely, to the extreme positive direction while it spends much more time mingling with mildly negative values. In other words, a positive skewness implies that most values are concentrated below the mean (which is zero for the case shown), with extreme values way above the mean.

The *fourth central moment* is defined as

$$\overline{u^4} \equiv \int_{-\infty}^{\infty} u^4 f(u)\, du \qquad (3.21)$$

Normalizing the fourth central moment with σ^4 gives us the kurtosis or flatness factor

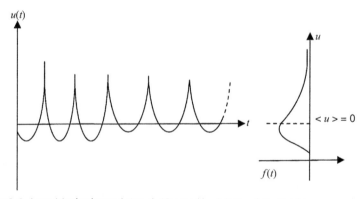

Figure 3.6 A positively skewed signal. *(Created by A.R. Vasel-Be-Hagh).*

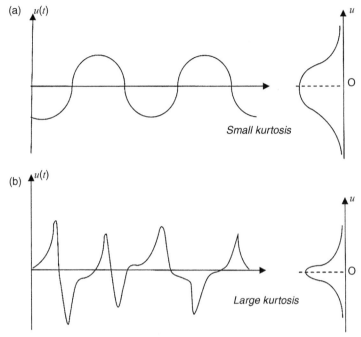

Figure 3.7 (a) A signal with a small kurtosis, (b) A signal with a large kurtosis. *(Created by A.R. Vasel-Be-Hagh).*

$$K \equiv \frac{\overline{u^4}}{\sigma^4} = \frac{1}{\sigma^4} \int_{-\infty}^{\infty} u^4 f(u)\, du \qquad (3.22)$$

Figure 3.7 shows the difference between a small kurtosis and a large kurtosis signal. We see that the larger the K, the flatter and wider the two tails (and the narrower the zero peak). For a normal (Gaussian) distribution, the flatness factor K of the Gaussian function is equal to three (wikipedia, 2015). One extreme is the discrete distribution with two equally probable outcomes, such as the tossing a coin where the outcome is either heads or tails. The flatness factor for this discrete distribution is unity (Brown, 2011). At the other extreme is the Student's t distribution, which has a flatness of infinity (Brown, 2011). Whatever the value may be, kurtosis or flatness factor connotes an important characteristic of the involved signal as portrayed in Fig. 3.8.

Figure 3.9 shows an actual PDF obtained from the turbulent flow generated by an orificed, perforated plate (Liu and Ting, 2007; Liu et al., 2007); more will be discussed in the chapter on grid turbulence. In a nutshell, the

Figure 3.8 Small versus big kurtosis. *(Created by S.P. Mupparapu).*

wavelet decomposition is simply a technique used to separate the cascade of eddying turbulent motions into bins of frequencies. The larger eddies are associated with the lower frequencies, and the smaller ones have higher frequencies. Figure 3.9 clearly portrays increasing flatness with decreasing wavelet (decomposition) level. In other words, the value of the flatness factor K associated with the large, low frequency, eddies are close to three, that is, near Gaussian, and it increases (the curve becomes more narrow and peaked) with increasing frequencies or decreasing eddy size. The increase in the kurtosis factor as we move toward the smaller eddies seems to suggest that the high-intensity

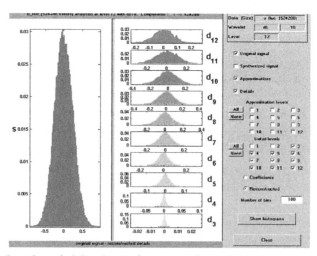

Figure 3.9 Sample probability density function and wavelet decomposition of orificed, perforated plate turbulence. *(Created by R. Liu).*

portion of the smaller eddying motion is rather intermittent. More will be said on this type of flow turbulence in the latter part of this book.

3.4 JOINT STATISTICS AND CORRELATION FUNCTIONS

In this section, we expand the discussion on statistical flow turbulence to consider both the x and y components simultaneously. As the eddying motions in turbulence involve finite volumes of the continuum fluid, we expect some correlation between the two components of velocity. In reality, the three orthogonal components are related; nevertheless, we will limit it to two components while noting that the discussion and analysis can be extended to all three dimensions. Let us denote the fluctuating velocity in the x direction as $u(t)$ and that in the y direction as $v(t)$. A sample time trace of each of these fluctuating velocities with zero mean is sketched in Fig. 3.10. The joint PDF $f(u,v)$ is proportional to the fraction of time that the two

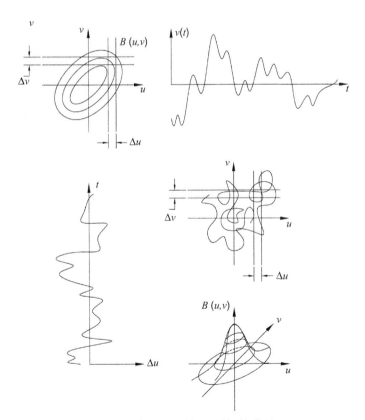

Figure 3.10 The joint probability function. *(Created by N. Cao).*

fluctuating components spend in a small window defined by u and $u + \Delta u$ and v and $v + \Delta v$. The time fraction cannot be negative, and the time spent at all locations must equal the total time, that is

$$f(u,v) \geq 0, \quad \int_{-\infty}^{\infty}\int_{-\infty}^{\infty} f(u,v)\,du\,dv = 1 \tag{3.23}$$

We note that summing all of the values of u at a given value of v gives us the probability density function of $u(t)$ at that v value. In other words, cutting a slice at $v = v_1$ yields the corresponding PDF $f(u$ at $v = v_1)$. Similarly, if all of the values of v at a given value of u are combined, we should get the PDF of $v(t)$. In short

$$\int_{-\infty}^{\infty} f(u,v)\,dv = f_u(u), \quad \int_{-\infty}^{\infty} f(u,v)\,du = f_v(v) \tag{3.24}$$

In flow turbulence, the most important joint moment is

$$\overline{uv} \equiv \int_{-\infty}^{\infty}\int_{-\infty}^{\infty} uvf(u,v)\,du\,dv \tag{3.25}$$

This is called the covariance or correlation between u and v. Figure 3.11a depicts a pair of negatively correlated random variables, while Fig. 3.11b corresponds to a positively correlated pair. If $\overline{uv} = 0$, $u(t)$ and $v(t)$ are said to be uncorrelated. Uncorrelated variables such as that shown in Fig. 3.11b, however, are not necessarily independent of each other (Tenneskes and Lumley, 1972). Two variables such as the x and y components of the turbulence fluctuations, as shown in Fig. 3.11d, are statistically independent if $f(u,v) = f_u(u)\,f_v(v)$; in which case, the probability density of one variable is not affected by the other variable, and vice versa. The joint characteristic function is the two-dimensional Fourier transform of the joint density, $f(u, v)$.

A sample plot of the covariance $\overline{uv}$ of turbulence created with an orificed, perforated plate (Liu and Ting, 2007; Liu et al., 2007) is plotted in Fig. 3.12. It is clear that for the "simple" turbulence generated by the orificed, perforated plate, $u(t)$ and $v(t)$ are uncorrelated, that is, $\overline{uv} = 0$. Furthermore, the shape is very close to that of Fig. 3.11d; thus, it appears that the turbulence generated by our orificed, perforated plate is very clean and the probability density of u is not affected by that of v, and vice versa.

The correlations discussed earlier deal with different components of the fluctuating velocity, which are synchronized or time-stamped. We now extend the statistical correlations to the case that deals with temporal variation,

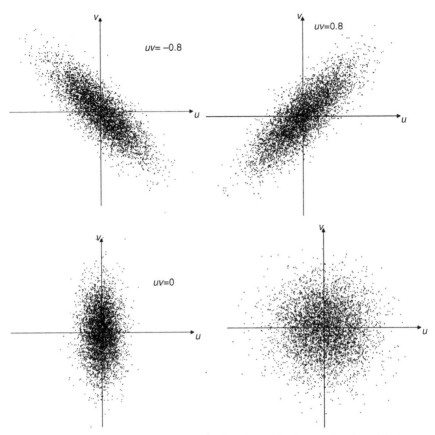

Figure 3.11 Negatively correlated, uncorrelated, positively correlated, and independent variables. *(Created by R. Liu).*

called joint moments and covariance. Let us consider two distinct times, t and t', and form the ensemble average of the product $u(t)\,u(t')$ for each realization. The corresponding joint moment, the covariance R_{uu}

$$R_{uu}(t,t') = \langle u(t)u(t')\rangle \tag{3.26}$$

Note the later time $t' = t + \tau$, where t is the reference time and τ is the delay. Since this deals with the same velocity component and x-fluctuating component u has been used as a generic case, the correlation $\overline{u(t)u(t')}$ is also called the autocorrelation; that is, correlating with itself, as the time difference between them varies. For "stationary" flow turbulence, the fluctuating velocity proceeds in time with homogeneity and hence, its statistical properties do not change with respect to time. In other words, the joint

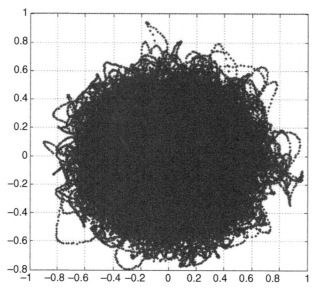

Figure 3.12 Sample covariance $\overline{uv}$ of orificed, perforated plate turbulence. *(Created by R. Liu).*

moments of stationary turbulence, such as the covariance defined in Eq. (3.26), are independent of the choice of time origin.

It is worth noting that the covariance $R_{uu}(\tau)$ typically decreases rapidly with increasing time difference. Furthermore, the autocovariance $R_{uu}(\tau)$ is an even function since

$$R_{uu}(\tau) = \langle u(t)u(t+\tau) \rangle = \langle u(t'-\tau)u(t') \rangle = R_{uu}(-\tau) \qquad (3.27)$$

In other words, because $\overline{u(t)u(t')} = \overline{u(t')u(t)}$, the autocorrelation is a symmetric function of τ.

Schwartz's inequality states that

$$\left| \overline{u(t)u(t')} \right| \leq \left[\overline{u^2(t) \cdot u^2(t')} \right]^{1/2} \qquad (3.28)$$

For stationary variable $u(t)$ associated with stationary flow turbulence, we have $\overline{u^2(t)} = \overline{u^2(t')}$, which is a constant. Therefore, we can define an autocorrelation coefficient $\rho(\tau)$ as

$$\frac{\overline{u(t)u(t')}}{\overline{u^2}} \equiv \rho(\tau) = \rho(-\tau) \qquad (3.29)$$

We can see that

$$|\rho| \le 1 = \rho(0) \tag{3.30}$$

The integral time scale τ_Λ is defined by

$$\tau_\Lambda \equiv \int_0^\infty \rho(\tau)\,d\tau. \tag{3.31}$$

This, along with the Taylor microscale τ_λ, is depicted in Fig. 3.13a. The value of τ_Λ is a measure of the temporal interval over which $u(t)$ is correlated with itself. Typical plots of the streamwise autocorrelation function, $f(r)$, where r is the spatial distance, at 20 diameters downstream of our orificed, perforated plate at 5.8, 7.8, and 10.8 m/s wind are shown in Fig. 3.13b.

The Taylor microscale τ_λ is defined by the curvature of the autocorrelation coefficient at the origin (Taylor, 1935, 1936); that is

$$\left.\frac{d^2\rho}{d\tau^2}\right|_{\tau=0} \equiv -\frac{2}{\tau_\lambda^2} \tag{3.32}$$

Expanding the autocorrelation coefficient ρ in a Taylor series about the origin, we can write, for small τ

$$\rho(\tau) \approx 1 - \tau^2/\tau\lambda^2 \tag{3.33}$$

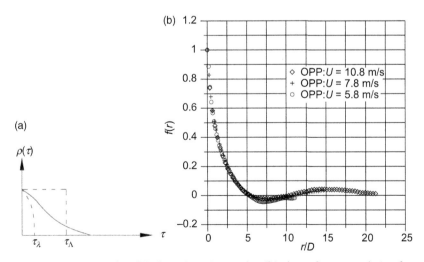

Figure 3.13 (a) Integral and Taylor micro time scales, (b) plots of autocorrelation function at 60D downstream of an orificed, perforated plate. *(Created by F. Iakovidis and R. Liu).*

That is, the microscale is the intercept of the parabola that matches $\rho(\tau)$ at the origin. Because $u(t)$ is stationary, we can write

$$0 = \frac{d^2}{dt^2}\left(\overline{u^2}\right) = \frac{d}{dt}\left(2\overline{u}\frac{d\overline{u}}{dt}\right) = 2\overline{\left[\frac{d\overline{u}}{dt}\frac{d\overline{u}}{dt} + \overline{u}\frac{d^2\overline{u}}{dt^2}\right]} = 2\overline{\left(\frac{du}{dt}\right)^2} + 2\overline{u\frac{d^2u}{dt^2}} \quad (3.34)$$

From Eq. (3.34), we obtain

$$\overline{\left(\frac{du}{dt}\right)^2} = \frac{2\overline{u^2}}{\tau_\lambda^2} \quad (3.35)$$

Another correlation is that associated with the variable and its time derivative. The cross-covariance of $u(t)$ and its time derivative $du(t + \tau)/dt$ is

$$R_{u\frac{du}{dt}}(\tau) = \left\langle u(t)\frac{du(t+\tau)}{dt}\right\rangle = \frac{\partial}{\partial\tau}\langle u(t)u(t+\tau)\rangle = \frac{\partial}{\partial\tau}R_{uu}(\tau) \quad (3.36)$$

We see that this autocorrelation of du/dt can be related to the autocorrelation coefficient ρ as

$$\overline{\frac{du(t)}{dt}\frac{du(t')}{dt}} = \overline{u^2}\frac{d^2}{dtdt'}\rho(t'-t) = -\overline{u^2}\frac{d^2\rho}{d\tau^2} \quad (3.37)$$

Moreover, for a stationary process such as stationary turbulence, the joint covariance function between the x component and the y component is

$$R_{uv}(\tau) = \langle u(t)v(t+\tau)\rangle \quad (3.38)$$

In general

$$R_{uv}(\tau) = R_{vu}(-\tau) \quad (3.39)$$

3.5 ADDITIONAL CONSIDERATIONS

At this point, it is appropriate to make a couple of comments concerning statistical analyses that have been covered in this chapter. The first is regarding the convergence of averages. In practice, we can only integrate over a finite time interval, that is

$$\overline{U}_T = \frac{1}{T}\int_t^{t+T} U(t)\,dt \quad (3.40)$$

The difference between this and the true mean is

$$\overline{U_T} - \overline{U} = \frac{1}{T}\int_0^T \left[U(t) - \overline{U} \right] dt = \frac{1}{T}\int_0^T u(t)\, dt \tag{3.41}$$

Here we took $t = 0$ for the sake of convenience. We note that the error tends to become smaller as the integration time increases and that the mean value found this way should stabilize to a constant value. Ergodicity is the requirement that a time average should converge to a mean value.

An ergodic variable is found when averages of all possible quantities formed from it converge. We note that a random variable becomes uncorrelated with itself at large time differences; the time difference τ approaches infinity, and it also becomes statistically independent of itself. For example, the integral time scale t_Λ of $u(t)$ is not only a measure of time over which $u(t)$ is correlated with itself, but also a measure of the time over which it is dependent on itself. When dealing with digital data, sampling once every two integral time scales is adequate to satisfy the Nyquist theorem; that is, the rate of data acquisition or sampling should be at least twice the maximum frequency of interest.

3.5.1 Fourier Series and Coefficients

Let us briefly recap Fourier series. The signal such as the fluctuating velocity in the streamwise direction

$$u(t) = A_0 + \sum_{n=1}^{\infty} \left(A_n \cos nt + B_n \sin nt \right) \tag{3.42}$$

is a periodic function with a period $T = 2\pi$. The Fourier coefficients are shown as

$$A_0 = \frac{1}{2\pi}\int_{-\pi}^{\pi} u(t)\, dt \tag{3.43}$$

$$A_n = \frac{1}{\pi}\int_{-\pi}^{\pi} u(t)\cos(nt)\, dt \tag{3.44}$$

$$B_n = \frac{1}{\pi}\int_{-\pi}^{\pi} u(t)\sin(nt)\, dt \tag{3.45}$$

The trigonometric series corresponding to $u(t)$ is called the Fourier series for $u(t)$. Note that the Fourier series, Eq. (3.42), can also be expressed as

$$u(t) = A_0 + \sum_{n=1}^{\infty} C_n \cos\left(n\omega t - \varphi_n\right) \tag{3.46}$$

or

$$u(t) = A_0 + \sum_{n=1}^{\infty} C_n \sin\left(n\omega t + \varphi_n^*\right) \tag{3.47}$$

where

$$C_n = \sqrt{A_n^2 + B_n^2} \tag{3.48}$$

$$\tan\varphi_n = B_n / A_n \tag{3.49}$$

$$\tan\varphi_n^* = A_n / B_n \tag{3.50}$$

For any nondeterministic waveform such as a real turbulent velocity, we can approximate the signal as

$$u(t) \approx A_0 + \sum_{n=1}^{\infty} A_n \sin\left(n\omega t + \varphi_n^*\right) \tag{3.51}$$

3.5.2 Fourier Transforms and Characteristic Functions

When the probability density undergoes Fourier transformation, we get the characteristic function. One of the applications of the characteristic function is to test whether the distribution of a random variable is Gaussian.

If we choose the PDF $f(u)$ and the corresponding characteristic function $\psi(k)$ as an example, a Fourier-transform pair can be defined as

$$\psi(k) \equiv \int_{-\infty}^{\infty} e^{iku} f(u)\,du, \quad f(u) \equiv \frac{1}{2\pi} \int_{-\infty}^{\infty} e^{-iku} \psi(k)\,dk \tag{3.52}$$

This gives

$$\psi(k) = \overline{\exp[\hat{i}ku(t)]} \tag{3.53}$$

In other words, $\psi(k)$ can be measured by averaging the output of a function generator that converts $u(t)$ into $\sin[u(t)]$ and $\cos[u(t)]$. Note that the convergence of $\psi(k)$ is much better than that of $f(u)$, because one must wait a long time to obtain a stable average.

Also note that averaging a function is equivalent to selecting the value of its Fourier transform at the origin. If the physical variable is time, the transform variable is frequency; the origin in transform space corresponds to zero frequency. For example, when we average a random variable, the only thing left is the component at zero frequency; all other components become zero.

Readers wishing to move further along this topic are recommended to consult Tennekes and Lumley (1972). And for those more mathematically and statistically versed, the two classic volumes by Monin and Yaglom (1971) may also be of interest.

Problems

Problem 3.1 Covariance

As far as the covariance $\overline{uv}$ is concerned, Fig. 3.11 shows negatively correlated, uncorrelated, positively correlated, and independent variables. Show, and if possible prove, using existing or artificially generated signals that for a case with $\overline{uv} = 0$ where $u(t)$ and $v(t)$ are dependent on each other. Generate another case where the two velocity components are independent of each other.

Problem 3.2 Correlation function and spectra

A cosine wave with an amplitude of unity and a frequency of 10 Hz has been recorded at a high sample rate (say, 20 kHz) over a moderately long time (say, 60 s). Estimate the integral and (Taylor) microscales from the autocorrelation.

Problem 3.3 Statistical analysis of partial grid turbulence

The velocity downstream of a partial grid installed in the middle of a wind tunnel test section as shown in Fig. 3.14 is measured using a 1-d hot wire probe. Data file P3-3 consists of 5000 velocity data points sampled at 80 kHz at $x/D = 10$, $y/D = 2$, and $z = 0$.

Part I. Plot the fluctuating velocity time trace.

Part II. Deduce the average velocity and comment on the effect of sample size on the value, if any.

Part III. Plot the probability density function and comment on the distribution.

Part IV. Calculate the second, third, and fourth central moments. Discuss the results.

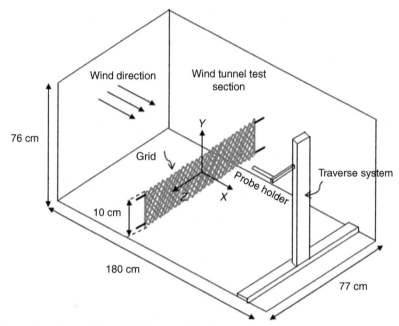

Figure 3.14 A partial grid in the middle of a wind tunnel test section. *(Created by F. Fouladi).*

REFERENCES

Batchelor, G.K., 1953. The Theory of Homogeneous Turbulence. Cambridge University Press, Cambridge.

Brown, S., 2011. Measure of Shape: Skewness and Kurtosis, http://www.tc3.edu/instruct/sbrown/stat/shape.htm (accessed 06.05.2015.).

Flierl, G., Ferrari, R., 2007. Turbulence in the Ocean and Atmosphere, Lecture #3: Statistics and Turbulence. MIT OpenCourseWare, http://ocw.mit.edu (accessed 09.05.2015.).

Garde, R.J., 2010. Turbulent Flow, third ed. New Age Science, Kent, UK.

Liu, R., Ting, D.S-K., 2007. Turbulent flow downstream of a perforated plate: sharp-edged orifice versus finite-thickness holes. J. Fluids Eng. 129, 1164–1171.

Liu, R., Ting, D.S-K., Checkel, M.D., 2007. Constant Reynolds number turbulence downstream of an orificed, perforated plate. Exp. Therm. Fluid Sci. 31, 897–908.

Monin, A.S., Yaglom, A.M., 1971. Statistical Fluid Mechanics: Mechanics of Turbulence, Volume I & Volume II, Dover New York.

Moon, F.C., 1992. Chaotic and Fractal Dynamics: An Introduction for Applied Scientists and Engineers, second ed. John Wiley & Sons, New York.

Taylor, G.I., 1935. Statistical theory of turbulence Parts 1–4. Proc. R. Soc. Lond. Ser. A 151, 421–478.

Taylor, G.I., 1936. Statistical theory of turbulence Part 5. Proc. R. Soc. Lond. Ser. A 156, 307–317.

Tennekes, H., Lumley, J.L., 1972. A First Course in Turbulence. MIT Press, Cambridge.

Wikipedia, 2015. http://en.wikipedia.org/wiki (accessed 02.05.2015.).

CHAPTER 4

Turbulence Scales

Genius is one percent inspiration and ninety-nine percent perspiration.
– Thomas A. Edison

Contents

Chapter Objectives

- To apply dimensional analysis to deduce key scales associated with laminar and turbulent flows.
- To discern and differentiate turbulent diffusivity from molecular diffusivity.
- To learn about Kolmogorov dissipative scales.
- To relate the Kolmogorov dissipative scales with the large inviscid eddies which supply the turbulent kinetic energy.
- To complete the key scales along the turbulent energy cascade, from energy-supplying large scales to energy-dissipating micro-scales.
- To introduce the turbulent kinetic energy spectrum.

NOMENCLATURE

a Acceleration
C Constant

Basics of Engineering Turbulence
http://dx.doi.org/10.1016/B978-0-12-803970-0.00004-0

c_p Heat capacity at constant pressure
D Diameter
E Energy
f Frequency
F Force
k Thermal conductivity
K Eddy diffusivity, exchange coefficient
ke   Kinetic energy per unit mass
l, L A characteristic length, a length scale
m Mass
P Pressure
P_{turb} Turbulence production rate (per unit mass of fluid)
q Total turbulence fluctuation from all directions
Re   Reynolds number, inertia force/viscous force, $Re = UD/v$
t A characteristic time period, a timescale
u Fluctuating velocity (in the x direction)
U Velocity
$\bar{U}$ Time-averaged (mean) velocity
v Fluctuating velocity in the y direction
x Distance along the x (streamwise) coordinate
y Distance along the y (vertically up) coordinate
z Distance along the z coordinate

Greek Symbols

α Thermal expansion coefficient
δ Boundary layer thickness
η Kolmogorov scale
θ Temperature (difference)
κ_t Thermal diffusivity, $\kappa_t = k/\rho c_p$
κ_w Wavenumber (cycles/m)
Λ Large length scale; integral length
λ Dissipative length; Taylor microscale
λ_w Wavelength (m/cycle)
μ Dynamic (absolute) viscosity
v Kinematic viscosity
ξ Small displacement
ρ Density
τ Shear; time scale
ε Dissipation rate
$\forall$ Volume

4.1 INTRODUCTION

... big whirls have little whirls that feed on their velocity, and little whirls have lesser whirls and so on to viscosity.

– Richardson, 1922

Richardson's legendary vision in 1922, as portrayed poetically above, is a profound breakthrough in our perception of flow turbulence. It suggests that the arcanum of flow turbulence is hidden in the ever-perplexing whirling motions that we call eddies. If this is true, then our scrutinizing eyes need to be focused on the presumably continuous cascade of eddying motions.

We limit the scope, unless otherwise explicitly stated, to fully developed turbulent flows where the energy cascade introduced by Richardson is well-defined. It is true that most turbulent flows in engineering applications are of significantly smaller Reynolds numbers than those encountered in the atmospheric flows which prompted Richardson's vision; nonetheless, the eddying cascade in many engineering applications is relatively well-defined. For these fully developed turbulent flows, the largest eddies, which are created by instabilities in the mean flow, are themselves subject to inertia instabilities, and thus rapidly break up and/or evolve into progressively smaller vortices. Dissipation is particularly pronounced in regions where the instantaneous gradient in velocity, and hence the shear stress, is large. Therefore, the dissipation of mechanical energy within a turbulent flow is concentrated in the smallest eddies. One may ask if the reverse process of smaller eddying motions converging into larger ones occurs. As can be inferred from Lim (1989) and Lim and Nickels (1992), this is definitely possible, although improbable. According to the second law of thermodynamics, all processes proceed from order to disorder. Therefore, even though there may be some eddies converging into more organized, larger ones, the overall trend is large eddies breaking down into smaller ones.

In this chapter, we will follow Wilson (1989) by first invoking the general scaling analysis on laminar and turbulent boundary layers in order to have a rudimentary knowledge of the possible players, that is, the scales involved. The crudely estimated turbulent diffusion is then cast in comparison to the molecular one associated with the fluid at the particular state; this shows the significant diffusion enhancement provided by flow turbulence. At that point, the dimensionally rigorously derived Kolmogorov scales are introduced, and thus we encounter the dustbin where all the turbulent kinetic energy is converted into heat. With the well-defined (at least in theory, irrespective of if there is truly such a physical scale) Kolmogorov scale, the mostly inviscid large eddies which supply the kinetic energy are estimated. With the two ends of the turbulence cascade prescribed, the key elements along the cascade are introduced and subsequently refined. The chapter ends with a brief introduction of the turbulent kinetic energy spectrum.

4.2 VELOCITY AND KEY LENGTH SCALES IN LAMINAR AND TURBULENT BOUNDARY LAYERS

In this section, a rudimentary dimensional approach is invoked to reveal the most apparent length, velocity, and time scales in the laminar and turbulent boundary layers. For the turbulence case, a large and a small length scale are introduced to crudely represent Richardson's turbulent energy cascade. Before we embark on this, let us express the Reynolds number in terms of diffusion time and advection time.

Figure 4.1 shows that the viscous diffusion from A to B separated by a distance L has a characteristic viscous diffusion time, t_v. This viscous diffusion time may be viewed as the time it takes B to feel the passing of A via fluid viscosity. We note that the viscous diffusion time is:

1. proportional to the distance; that is, the farther apart they are, the longer time it takes for B to feel the passing (effect) of A

$$t_v \sim L \tag{4.1}$$

2. inversely proportional to the fluid viscosity; that is, the larger the viscosity, the shorter the time it takes for B to feel A (and vice versa)

$$t_v \sim 1/v \tag{4.2}$$

Hence, dimensionally, this characteristic viscous diffusion time which is required for momentum to diffuse a distance L due to viscosity is

$$t_v = L^2/v, [m^2/(m^2/s)] = [s] \tag{4.3}$$

where v is the kinematic viscosity and square brackets [] enclose the units. Here we recall that the kinematic viscosity, $v = \mu/\rho$ have units $[N \cdot s/m^2]/[kg/m^3] = [m^2/s]$, where μ is the dynamic (absolute) viscosity, and that for a Newtonian fluid, the shear stress, $\tau = \mu \, dU/dy$, that is, $\mu = \tau/(dU/dy)$.

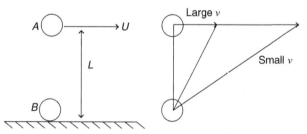

Figure 4.1 Diffusion time. *(Created by B. Motameni).*

Another common time scale is the advection time. For a body of length L in a flow field with a mean velocity U, a characteristic (overall) advection time scale t_a signifies the duration over which the fluid element is of significance to the body and vice versa. The advection time

$$t_a = L/U \qquad (4.4)$$

is the time it takes a fluid element to pass a body of length L. It can be regarded as the time required for momentum to advect across a distance L.

The foremost important nondimensional parameter in fluid mechanics is the Reynolds number. It signifies the strength of the inertia force with respect to the underlying viscous force of a moving fluid

$$\text{Re} = \text{inertia force}/\text{viscous force} = \text{advective}/\text{viscous effect} = (1/t_a)/(1/t_v) \quad (4.5)$$

We note that the shorter the time the fluid (element) takes to pass (advect) a distance L, the larger the inertia force. In other words, the faster the advection of a fluid particle is, the larger its inertia, and thus, the larger the corresponding Reynolds number. We can recast the Reynolds number expression as

$$\text{Re} = t_v/t_a = (L^2/v)/(L/U) = U L/v \qquad (4.6)$$

where at room temperature and pressure, the kinematic viscosity $v \approx 1.5 \times 10^{-5}$ m^2/s for air, and for water, $v \approx 1.1 \times 10^{-6}$ m^2/s. This expression can also be derived from Newton's second law[1]. We sense that when the Reynolds number is small, there is enough viscous force to take care of the agitative inertia force and hence, turbulence is under control. With an increasing Reynolds number, the inertia force increases until the viscous force becomes incapable of keeping the fluctuations under control and the flow becomes turbulent. This is analogous to having a flow turbulence class with a large group of sugar-buzzed students. The professor does not have enough viscosity to calm the class down. Filling up the rowdy classroom with a high-viscosity fluid such as cold honey would presumably solve the challenge at hand. With this appreciation of viscosity diffusion, advection and inertia, along with Reynolds number, let us press on and explore some key scales involved in the ever-important boundary layer.

---

[1] According to Newton's second law, $F = ma = m(UdU/dx)$, dividing by volume $\forall$, we have $F/\forall = ma/\forall = (\rho\forall)(U^2/L)/\forall = \rho U^2/L$. Substituting this (inertia) force per unit volume into Re, we get $\text{Re} = (\rho\, U^2/L)/(\mu\, U/L^2)$ where the denominator is viscous force per unit volume. This can be simplified into $\text{Re} = \rho\, U L/\mu = U L/v$.

4.2.1 Laminar Boundary Layer

As detailed in Chapter 2, the Navier-Stokes equations for steady (laminar) flow of an incompressible fluid with constant viscosity can be expressed as

$$U_j \frac{\partial U_i}{\partial x_j} = \frac{1}{\rho} \frac{\partial P}{\partial x_i} + \nu \frac{\partial^2 U_i}{\partial x_j \partial x_j} \tag{4.7}$$

Let us consider the simple case of an initially uniform velocity flow over a flat plate as depicted in Fig. 4.2. For the laminar boundary layer flow, we may estimate the inertia terms on the left of Eq. (4.7) as U^2/L and the viscous terms signified by the last term as $\nu U/L^2$. The ratio of these terms is the Reynolds number, $Re = UL/\nu$. We see that viscous terms should become negligible, and thus may be dropped, at large Re. However, boundary conditions or initial conditions may make it impossible to neglect viscous terms everywhere in the flow field. For example, the viscous terms cannot be neglected in the velocity boundary layer, first introduced by Prandtl, as per brief historic account in the forthcoming paragraph and as portrayed in Fig. 4.2. This is particularly true in the inner portion of the boundary layer.

For many years, as jested by the British chemist and Nobel laureate Sir Cyril Norman Hinshelwood (1897–1967), fluid dynamists were divided into groups. On one end, there were those practical and applied hydraulic engineers who observed things that could not be explained. At the other end, there were those theoretical and pure mathematicians who explained things that could not be observed. This dilemma continued until the German engineer Ludwig Prandtl gave an epoch-making presentation to the Third International Congress of Mathematics held at Heidelberg; see, for example, Gad-el-Hak (1998). Prandtl (Prandtl, 1904; Anderson, 2005) introduced the concept of a fluid boundary layer adjacent to the surface of a moving body where viscous forces are important and outside of which the flow is more or less inviscid.

When considering length scales, we tend to associate viscous effects with small length scales. In other words, the viscous terms can survive at high Re only by choosing a length scale δ, which represents the thickness of the boundary layer as shown in Fig. 4.2, such that the viscous terms are of the same order of magnitude as the inertia terms; that is, from Eq. (4.7), we have

$$U^2/L \sim \nu U/\delta^2 \tag{4.8}$$

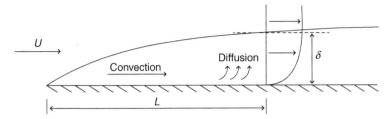

Figure 4.2 Laminar boundary layer. *(Created by B. Motameni).*

The viscous length δ is thus related to the scale L of the flow field as

$$\delta/L \sim [\nu/(UL)]^{\frac{1}{2}} = \mathrm{Re}^{-\frac{1}{2}} \tag{4.9}$$

We note that in shear flows, a "diffusive" length scale is associated with the diffusion across the flow, while a "convective" length scale is associated with the convection along the flow.

We can apply the asymptotic approximation to see what happens when δ/L approaches zero. At this limit, the shear flow becomes independent of most of its environment, except for the boundary conditions imposed by the overall flow. In other words, for large Reynolds number flows, viscosity is largely nonexistent except at the very thin layer next to a solid surface.

4.2.2 Turbulent Boundary Layer

Let us reconsider the boundary-layer flow for the case where the boundary layer is turbulent; see Fig. 4.3. The considerable inertia associated with large Re produces a lot of eddying motions. These turbulent eddies transfer momentum deficit away from the solid surface, making the cross-stream diffusion turbulent. The boundary-layer thickness δ presumably increases roughly as $d\delta/dt \sim u$, or $d\delta/dx \sim u$. In other words, the higher the turbulence level, the faster the boundary layer increases with respect to time or distance downstream. The time interval that has elapsed for a fluid particle moving from $x = 0$ to L (distance downstream as shown in Fig. 4.3) is on the order L/U, which is the convective time scale. Hence, we have

$$\delta \sim ut \sim u(L/U) \tag{4.10}$$

In effect, we are equating the turbulent diffusion time scale δ/u to the convective time scale L/U. We can rewrite the relation as $\delta/L \sim u/U$ or $\delta/u \sim L/U$.

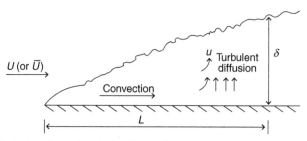

Figure 4.3 Turbulent boundary layer. *(Created by B. Motameni).*

The above analysis implies that with an imposed external flow the turbulence must have a time scale commensurate with the time scale of the flow. In other words, the flow turbulence is a function of the mean flow, entailing that turbulence is part of the flow and not part of the fluid as molecular diffusivity and viscosity are. Not all of the turbulence, however, has such a large time scale. The small eddies in turbulence have short time scales, which tend to make them statistically independent of the mean flow; that is, the smaller turbulent eddies may not depend directly on the mean flow. Therefore, more than one eddy size is needed to describe a turbulent flow. At a minimum, we need to introduce both a large length scale (as done earlier) and a small length scale to approximate flow turbulence.

It is known that for turbulent flows, the fluid motion creates a multitude of eddies that are responsible for transport properties. The molecular effects in turbulent flows act mainly to provide a sink for dissipation of small-scale eddying motion and to transport heat, mass, and momentum over distances less than that of the smallest turbulent eddies. In other words, a wide range of length scales exists in a turbulent flow, bounded by dimensions of the flow field and/or the body generating the flow disturbance and the diffusion action of molecular viscosity (molecular mean free path). Hopefully, there are some relations between the various scales of the mean motion and those of the turbulent eddies. These scale relationships will allow us to develop general predictions of the changes that occur in turbulence structure when the mean field is altered. Because eddies come in a wide range of sizes, a minimum of two length scales are needed to characterize the large and small eddies. Following the typical convention, the scale "Λ" is used to signify the large eddies and "λ" for the small eddies, which act as a sink for molecular motions to dissipate turbulence, channeling the turbulent kinetic energy via viscosity into heat; see Fig. 4.4.

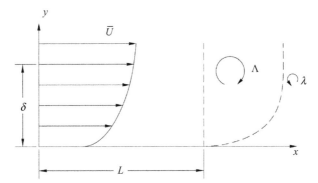

Figure 4.4 Large and small eddies in a turbulent boundary layer. *(Created by N. Cao).*

Recall that for a laminar flow, we have U as the velocity scale, and δ and L as the cross-stream and streamwise length scales, respectively. For a turbulent (boundary-layer) flow, we have $\bar{U}$, δ, and L signifying the mean (time-averaged) velocity and length scales; the corresponding fluctuating turbulence scales are u, Λ, and λ (see Fig. 4.4). The time scales for the mean flow and for the large scales turbulence are hence $L/\bar{U}$ and Λ/u, respectively.

4.3 MOLECULAR VERSUS TURBULENT DIFFUSION

Consider a stagnant fluid sandwiched between two solid boundaries as portrayed in Fig. 4.5. Suppose the floor is heated, but the fluid remains macroscopically stagnant. As such, the thermal energy is distributed via molecular diffusion, that is

$$\frac{\partial \theta}{\partial t} = \kappa_t \frac{\partial^2 \theta}{\partial x_i \, \partial x_i} \tag{4.11}$$

where θ is the temperature and the thermal diffusivity κ_t as a first approximation can be assumed to be constant. Dimensionally, this may be interpreted as

$$\frac{\Delta \theta}{t_{mol}} \sim \kappa_t \frac{\Delta \theta}{L^2} \tag{4.12}$$

where t_{mol} is the molecular diffusion time and $\Delta \theta$ is a characteristic temperature difference. This can be rewritten as

$$t_{mol} \sim L^2 / \kappa_t \tag{4.13}$$

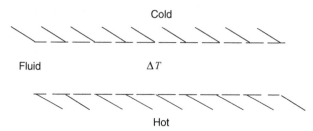

Cold

Fluid ΔT

Hot

Figure 4.5 Molecular versus turbulent diffusion. *(Created by D. Ting).*

If the ceiling is 2 m from the floor and the fluid is air which has a thermal diffusivity $\kappa_t = 0.20$ cm²/s at room temperature and pressure, then, $t_{mol} \sim 10^6$ s. In other words, in the absence of fluid motion, it will take days for the temperature to even out!

In reality when the fluid in the vicinity of the hot floor is heated, buoyancy-driven instabilities will emerge. If we assume that the resulting turbulent motion has a length scale on the order L and a velocity scale on the order u, then, the characteristic time is

$$t_{turb} \sim L/u \tag{4.14}$$

Let us further assume that the characteristic turbulence velocity u is on the order of cm/s. The resulting turbulence characteristic time for the thermal energy to be distributed throughout the compartment is then on the order of minutes. In other words, turbulence enhances the diffusivity by orders of magnitude, as compared to that involving molecular diffusivity alone.

We see from the above example that flow turbulence can boost the diffusivity by orders of magnitude. The seriously boosted diffusivity may be viewed as an effective diffusivity. By considering it in this way, we tend to treat turbulence as a property of a fluid rather than a property of a flow. This is somewhat dangerous conceptually, but it eases the analysis significantly and, by and large, it makes practical approximations versatile for many engineering applications.

Proceeding along the above assumption, we can express the diffusion of heat by turbulent motions as

$$\frac{\partial \theta}{\partial t} = K \frac{\partial^2 \theta}{\partial x_i \, \partial x_i} \tag{4.15}$$

where K is the "eddy diffusivity" or "exchange coefficient" for heat (thermal energy). The time scale according to this equation is $t \sim L^2/K$, but we also have $t_{turb} \sim L/u$ from before while relating turbulence with mean flow, and hence

$$t \sim L^2/K \sim L/u \qquad (4.16)$$

In other words, eddy diffusivity K is on the order of the product of turbulence fluctuating velocity and the characteristic length associated with the mean floor or physical dimension of the confinement, uL. This eddy diffusivity or viscosity K may be compared with the kinematic viscosity v and the thermal diffusivity κ_t

$$\frac{K}{\kappa_t} \cong \frac{K}{v} \sim \frac{uL}{v} = \mathrm{Re} \qquad (4.17)$$

This expression conveys the Reynolds number Re as the apparent or turbulent viscosity/molecular viscosity ratio. It is worth mentioning that the eddy diffusivity K is an artifice, which may not represent the effects of turbulence faithfully. With this caution in mind, this analogy between eddy diffusivity and the molecular diffusivity can resolve, though only as a first approximation, many practical problems.

4.4 KOLMOGOROV MICROSCALES OF DISSIPATION

We have so far discussed only the large scales Λ, along with a brief mention of some small scales λ. Let us press on to introduce the renowned Kolmogorov microscales of dissipation to "close the lid at the smallest eddies end". We note that large eddies do most of the transportation of momentum and contaminants and are generally the relevant length scales in the analysis of the interaction of turbulence with mean flow. The generation of small-scale fluctuations from the larger ones is due to the nonlinear terms in the equations of motion; that is, the viscous terms prevent the generation of infinitely small scales of motion by dissipating small-scale energy into heat.

One might expect that at large Re, the relative magnitude of viscosity is so small that the viscous effects in a flow tend to become vanishingly small. The nonlinear terms in the Navier-Stokes equation, however, counteract this threat by generating motion at scales small enough to be affected by viscosity. In other words, the smallest scale of motion automatically adjusts itself to the value of the viscosity.

Since small-scale motions tend to have small time scales, one may assume that these fast (high frequency) motions are statistically independent of the relatively slow large-scale turbulence and the mean flow. Alternatively, we can argue that the offspring of nonlinear interactions after passing down via generations of nonlinear interactions from the originally large and deterministic motions tend to forget their origins. This is somewhat analogous to the result of many generations of inter-racial mixing, that is, offspring with no distinctive ethnical origins. As byproducts of many nonlinear interactions, the small-scale motions may be perceived as a passive outcome[2] of the rate of energy supply via the large-scale motion. In addition, they are also directly affected by the kinematic viscosity of the fluid that they are in. Thus, small-scale motion is a sole function of the rate of energy supply by the large-scale motion and the fluid kinematic viscosity.

With the aforementioned view, the small eddies are formed due to the handing down of energy from the large eddies; the rate of energy supply (from the mean flow into the large eddies) is thus approximately equal to the rate of dissipation, ε. This is particularly true for "stationary" turbulence, which does not alter with time. For typically large Reynolds number turbulent flows, the rate of dissipation adjusts itself to the amount of energy funneling down the energy cascade; therefore, this quasi-equilibrium assumption is likely valid, provided the change is not too rapid. In other words, provided the net rate of change is significantly less than the rate of energy dissipation, we may assume Kolmogorov's universal equilibrium theory (Kolmogorov, 1941). One important outcome of the Kolmogorov's universal equilibrium theory is that the dissipation rate per unit mass ε (m^2/s^3) and the kinematic viscosity ν (m^2/s) govern the small-scale motion.

In other words, Kolmogorov (1941) developed a set of dissipation velocity and length scales that are independent of any large eddy turbulence properties. His two observations were:

1. Because the smallest eddies are dissipated by viscosity, the size of scale η necessary to carry out a fixed rate of dissipation should be a function of only viscosity ν;

---

[2] As mentioned earlier, this general trend of large to small eddying motions is consistent with the second law of thermodynamics. In other words, the original large eddies formed by mean flow are more or less organized and deterministic. In accordance with the second law of thermodynamics, they break down into progressively more disorganized and random smaller eddies. Note that this does not exclude some smaller eddies coming together and form larger ones, but the overall process is, by and large, from more organized larger eddies breaking into smaller ones.

2. Because the rate of dissipation is related to the size η of eddies for a fixed viscosity, this size η should be a function of only the dissipation rate ε.

Combining these ideas, Kolmogorov proposed that the length, velocity and time scales, η, u_η, and t_η of dissipation should be a function only of dissipation rate and viscosity. Note that this hypothesis can only be correct if the dissipation eddies are much smaller than those that participate in production. If this is not the case, eddies may be simultaneously involved in both production and dissipation and will have their length scales influenced by mean shear, in addition to dissipation rate and viscosity. As mentioned earlier, we limit the bulk of this book to fully developed turbulence where there is a well-defined cascade of turbulent eddies, from large energy-supplying eddies to small dissipative microscales. Under such a condition, the Kolmogorov hypothesis is expected to be viable.

Recall that the dimensions of dissipation and viscosity are m²/s³ and m²/s, respectively, and hence, via dimensional analysis

$$\eta \sim \varepsilon^a \, v^b \tag{4.18}$$

we find $a = -1/4$ and $b = 3/4$. The resulting length, velocity, and time scales are, respectively

$$\eta = (v^3/\varepsilon)^{1/4} \tag{4.19}$$

$$u_\eta = (\varepsilon_v)^{1/4} \tag{4.20}$$

$$t_\eta = (v/\varepsilon)^{1/2} \tag{4.21}$$

Liepmann (1979) gave a physical argument for the Kolmogorov scales. Knowing that turbulence can only exist at large Reynolds numbers, based on local eddy size and velocity, there must be an eddy scale η with its corresponding velocity u_η at which turbulence ceases. This occurs for a local Reynolds number of about unity, that is

$$\mathrm{Re}_\eta = u_\eta \eta/v \approx 1 \tag{4.22}$$

The Reynolds number, $\eta u_\eta/v \approx 1$ illustrates that the small-scale motion is quite viscous and that the viscous dissipation adjusts itself to the energy supply by adjusting the corresponding length scales. Specifically, the smallest eddy that survives long enough to be identified as an eddy is the one for

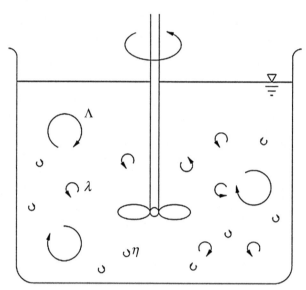

Figure 4.6 Kolmogorov and larger scales in a mixing process. *(Created by N. Cao).*

which the viscous diffusion time v/u_η^2 is approximately equal to the eddy advection time, η/u_η. We may view this approximate equality as an eddying fluid motion that dissipates completely, transforming all its kinetic energy via viscosity into heat within one rotation. Thus, the energy dissipation of these elements is purely viscous, and if all the energy produced from the mean flow is dissipated by these small scales, the overall dissipation rate ε is equal to the small scale rate and we can define

$$\varepsilon = vu_\eta^2/\eta^2 \tag{4.23}$$

Solving Eqs (4.22 and 4.23) simultaneously yields the Kolmogorov scales η and u_η as expressed earlier. We see that an increase in viscosity can enhance the dissipation, as expected. Increasing the velocity augments the shear and thus also the rate of dissipation. On the other hand, enlarging the eddy size leads to a reduced velocity gradient, and hence, a decrease in the dissipation rate.

Let us follow Tennekes and Lumley (1972) and consider a mixing process in which the mixture involved is a liquid having a viscosity $v = 10^{-3}$ m²/s and a density of approximately 1000 kg/m³; see Fig. 4.6. Suppose a 20 W electric mixer is used to mix a 1 L mixture. At equilibrium conditions the power input is equal to the dissipating power, that is, $\varepsilon = 20$ W/kg. The corresponding Kolmogorov length scale $\eta = (v^3/\varepsilon)^{1/4} = 2.7$ mm. We also note that halving of the Kolmogorov eddy size would require a power increase by a factor of 16. This is due to the ¼ exponent involved in the equation. Hence, there is a lot of room to dissipate an enormous amount of energy (power) input before the smallest dissipative turbulent eddies approach the molecular mean free path, at which point the continuum assumption becomes invalid.

4.5 AN INVISCID ESTIMATE FOR DISSIPATION RATE

If we can relate the dissipation rate ε with the length and velocity scales of the large-scale turbulence, we can form a good impression of the differences between the large-scale and small-scale aspects of turbulence. Let us invoke the assumption that the rate at which large eddies supply energy to small eddies is inversely proportional to the time scale of the large eddies. This may be interpreted as a large eddy passing all of its kinetic energy unto smaller eddies within its life span. And, if we follow along Richardson's energy cascade proposition, these smaller eddies subsequently pass all their energy unto even smaller eddies when making their revolution. This is quite similar to some well-off parents who pass all their fortune onto their children, who in turn hand it down to their offspring, and so on. To make the analogy more applicable, the amount of inheritance remains largely the same, and the number of offspring drastically multiplies down (or up if we consider an upright tree) the family tree. Let us start from the beginning (roots) of the ancestry where the large eddies are formed via shear in the mean flow, that is, by taking away some significant amount of energy from the main flow (they make a fortune from nothing but hard work). These eddies subsequently pass all that they have down to their immediate, multiple offspring as they pass away; in reality, they themselves break down into many smaller and faster-spinning eddies. With no net change in the total assets (kinetic energy), the amount per living member decreases drastically from one generation to the next because of the manifold procreation. Other than the increase in number and the corresponding decrease in size, the spending habit also escalates down the lineage, that is, unto the progressively faster whirling smaller eddies. The passing down of the said inheritance continues until the generation of eddies are so small and so hastily twirling (squandering) that they dissipate all their energy via viscosity into heat within their short life span, leaving neither energy nor offspring.

Returning back to the large, energy-containing eddies, the kinetic energy per unit mass is proportional to u^2; for $\frac{1}{2}mu^2/m \approx u^2$ (order of magnitude-ly speaking, we drop the "½"), where m is the mass. The rate of energy transfer is proportional to u/Λ (where Λ represents the size of the large eddies); that is, the energy transfer rate varies inversely with the time scale associated with the large eddies. Accordingly, we see that the energy supply rate is related to $u^2 \times u/\Lambda$, which is equal to the dissipating rate under quasi-equilibrium condition. Hence, following Taylor (1935) we have

$$\varepsilon \sim u^3/\Lambda \qquad (4.24)$$

which states that viscous dissipation can be estimated from the large-scale dynamics, which do not involve viscosity. In other words, dissipation is a passive process; that is, it proceeds at the rate dictated by the inviscid inertial behavior of the large eddies. Note that some recent studies have suggested that a significant nonequilibrium region can exist in some turbulent flows in which this expression may not hold true (Vassilicos, 2015). It is worth stressing that we are primarily concerned with equilibrium or quasi-equilibrium conditions where this order of magnitude equation is applicable.

The energy cascade described above is one of the cornerstone assumptions of turbulence theory, claiming that large eddies lose a significant fraction of their kinetic energy (per unit mass) ½ u^2 within one "turnover" time Λ/u. The nonlinear mechanism that produces small eddies out of larger ones is as "dissipative" as its characteristic time permits; that is, they "dissipate" (or break down) into progressively smaller eddies. Thus, turbulence is a strongly damped nonlinear stochastic system.

For large Re, large eddies lose a negligible fraction of their energy to viscous dissipation effects directly; indirectly, they lose a whole lot to smaller eddies. The time scale of their decay is relatively large at Λ^2/ν. We see that the larger the eddy is, the longer it lasts, and that the more viscous the fluid is, the shorter it lives. We further note that viscous energy loss (for eddies of u and Λ) proceeds at a rate $\nu u^2/\Lambda^2$, which is small compared to u^3/Λ for large Re ($u\Lambda/\nu$). In the **energy cascade**, kinetic energy is transferred to successively smaller and smaller eddies, until the Reynolds number of the eddy is sufficiently small that the eddy motion is stable and molecular viscosity is effective in dissipating the kinetic energy.

Based on the materials conveyed up to this point, we may relate the relatively well-defined Kolmogorov scales with the more approximately estimated scales associated with the large eddies. We see that:

1. $\eta/\Lambda = (\nu^3/\varepsilon)^{1/4}/\Lambda \sim (\nu^3/u^3/\Lambda^3)^{1/4} = (u\Lambda/\nu)^{-3/4}$, or

$$\eta/\Lambda \sim (u\Lambda/\nu)^{-3/4} = \mathrm{Re}^{-3/4} \qquad (4.25)$$

2. $t_\eta/t_\Lambda = (v/\varepsilon)^{\frac{1}{2}}/t_\Lambda \sim (v/\varepsilon)^{\frac{1}{2}}/(\Lambda/u) \sim (v/u^3/\Lambda)^{\frac{1}{2}}/$
$(\Lambda/u) = (v/u\Lambda)^{\frac{1}{2}} = (u\Lambda/v)^{-\frac{1}{2}}$, or $\qquad\qquad$ (4.26)

$$t_\eta/t_\Lambda \sim (u\Lambda/v)^{-\frac{1}{2}} = \mathrm{Re}^{-1/2}$$

3. $u_\eta/u_\Lambda = (v\varepsilon)^{1/4}/u_\Lambda \sim (vu^3/\Lambda)^{1/4}/u = (v/u\Lambda)^{1/4} = (u\Lambda/v)^{-1/4}$, or
$$u_\eta/u_\Lambda \sim (u\Lambda/v)^{-1/4} = \mathrm{Re}^{-1/4} \qquad\qquad (4.27)$$

These equations imply that the smallest eddies are significantly smaller than the largest ones, especially at larger Reynolds numbers. Second, the smallest time scales are much briefer than the largest time scales, and this difference increases with $\mathrm{Re}^{1/2}$. Third, the smallest velocity scales are lower than the largest velocity scales, and this ratio varies with $\mathrm{Re}^{1/4}$. Table 4.1 illustrates how these small-large length scale ratios vary with respect to the Reynolds number. It is clear that the separation in scales widens as Re increases, leading to progressively more evident small-scale structures. As the largest scales are predominantly set by the physical confinement or bluff body involved, they do not tend to noticeably alter with Reynolds number in the absence of changes to the physical dimensions. Under such (physical invariant) conditions, essentially only the passive, smaller-length scales are adjusted by variations in Reynolds number; that is, the smallest length drastically (to the power ¾ as per Eq. (4.25)) decreases, its time scale is significantly reduced (to ½ power according to Eq. (4.26)), and its velocity is lowered (to the power of ¼ as expressed by Eq. (4.27)), with increasing Re. For the example illustrated in Table 4.1, doubling the Reynolds number resulted in an approximately 40% decrease in η/Λ, 30% reduction in t_η/t_Λ, and near 20% attenuation in the corresponding velocity ratio u_η/u_Λ.

At this point, it is worth taking a look at the vorticity, which has the dimension of frequency $[\mathrm{s}^{-1}]$, associated with these large and small scales. The small-large scales vorticity ratio is

$$f_\eta/f_\Lambda \sim t_\Lambda/t_\eta \sim (v/u\Lambda)^{\frac{1}{2}} = \mathrm{Re}^{1/2} \qquad\qquad (4.28)$$

Table 4.1 Variations of small/large scales with Re

Re	5,000	10,000
η/Λ	0.0017	0.001
t_η/t_Λ	0.014	0.01
u_η/u_Λ	0.12	0.1

This clearly depicts that the vorticity of the small-scale eddies is much greater than that associated with the large-scale eddies. Hence, it is not surprising to see the use of the small, rather than the large, scale in modeling turbulence via vortex dynamics. As importantly, the scrutinizing eyes should focus on the small scales in a situation where the underlying vorticity may play a vital role in the problem at hand.

On the other hand, if we look at the corresponding distribution in the turbulent kinetic energy, we see that the mass of a Kolmogorov eddy $m_\eta \approx \rho\eta^3$, which is puny compared to that associated with a large eddy $m_\Lambda \approx \rho\Lambda^3$. In other words, the kinetic energy of the smaller eddies is substantially less than that of the large eddies; that is, $\frac{1}{2}\, m_\eta\, u_\eta^2 << \frac{1}{2}\, m_\Lambda\, u_\Lambda^2$. In summary, most of the fluctuating energy is associated with large-scale motions, while most of the vorticity is associated with small-scale motions. In engineering practice one should understand the problem at hand, that is, whether the kinetic energy or the vorticity plays a more important role. Then again, there are pieces of music where every instrument of the orchestra is necessary. In other words, we may not be able to dismiss the large or small scales in some situations; in fact, even the in-between scales are needed to accurately comprehend and "solve" many flow turbulence problems.

4.6 THE ENERGY CASCADE – SCALES FROM PRODUCTION-DISSIPATION ENERGY BALANCE

Thus far we have gathered some understanding of the energy cascade where kinetic energy is harvested from the mean flow by the large scales through velocity gradient or shear and is transferred down to successively smaller eddies until viscosity. Let us move forward and introduce another dissipative length construed by Taylor (1935) and develop some approximate expressions concerning the energy budget. We will take for granted the fundamental assumption that the turbulence is in equilibrium, with production from the Reynolds stress-mean shear interaction balanced by the continuous destruction of turbulence by viscous dissipation. The key point is that the eddies that produce most of the dissipation are much smaller than the eddies which contain Reynolds shear stresses $\overline{uv}$ that cause turbulence production[3]. This requires a reasonably large Reynolds number and/or fairly well-developed turbulence, a condition of primary interest of this

---

[3] In other words, large-scale turbulence fluctuations are generated by the mean flow via the Reynolds stresses.

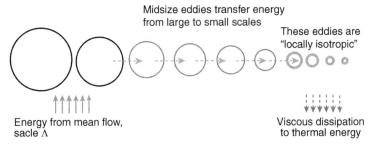

Figure 4.7 The turbulence energy cascade. *(Created by D. Ting)*.

book. Figure 4.7 shows an overall view of the energy cascade, from production to dissipation.

4.6.1 Production, Dissipation, and Local Equilibrium

Though limited to a one-dimensional impression, Fig. 4.7 depicts some sort of energy cascade where the large eddies take energy from the mean flow via the Reynolds stresses and feed it down a cascade of progressively smaller eddies; at the end of the cascade, the smallest eddies dissipate the kinetic energy into thermal energy. The local equilibrium assumption implies that the production of turbulence is equal to dissipation locally. This requires that the other turbulence transport terms balance each other.

The total turbulence fluctuation from the three orthogonal contributions in a Cartesian coordinate system, $\overline{q^2} = \overline{u^2} + \overline{v^2} + \overline{w^2}$, where u, v, and w are the fluctuating velocities in the x, y, and z directions, respectively. The total change in this turbulence kinetic energy per unit mass of fluid is

$$\frac{D}{Dt}\left(\frac{\overline{q^2}}{2}\right) = \left(\text{pressure and turbulence diffusion}\right) + \left(\text{viscous work}\right)$$
$$+ \left(\text{production}\right) - \left(\text{viscous dissipation}\right) \tag{4.29}$$

In the absence of pressure and turbulence diffusion and viscous work, we are left with the production and viscous dissipation terms. These production P_{turb} and dissipation ε rate per unit mass are, respectively

$$P_{\text{turb}} = -\overline{u_i u_j}\frac{\partial \overline{U}_i}{\partial x_j} \tag{4.30}$$

$$\varepsilon = 2 v s_{ij}\overline{\frac{\partial u_i}{\partial x_j}} \tag{4.31}$$

Substituting for the fluctuating stress tensor from Chapter 2, we get

$$\varepsilon = 2\nu \overline{\left[\frac{1}{2} \left(\frac{\partial u_i}{\partial x_j} + \frac{\partial u_j}{\partial x_i} \right) \right] \frac{\partial u_i}{\partial x_j}} \tag{4.32}$$

Note that P_{turb} and ε have the units (velocity)^3/length = [m^2/s^3]. Using Newton's second law, which states that force is equal to mass times acceleration, this can be written as m^2/s^3 = N·m/s·kg = W/kg, by relating force, mass and length through Newton = kg·m/s^2. In a boundary layer with a prevailing flow in the horizontal (x) direction (see Fig. 4.4), the vertical velocity gradient is dominant and the production term, Eq. (4.31), becomes

$$P_{turb} \approx -\overline{uv} \frac{\partial \overline{U}}{\partial y} \tag{4.33}$$

Taylor (1935) showed that for isotropic turbulence, the complicated dissipation expression, Eq. (4.32), could be reduced to

$$\varepsilon = 15\nu \overline{\left(\frac{\partial u}{\partial x} \right)^2} \tag{4.34}$$

Recall that dissipation is a passive process, which is dominated by small scale motions. These small scales evolve from generations of intermixing and thus have more or less lost their memory of their origin, that is, the original mean flow orientation. In short, small-scale motions are roughly isotropic for well-developed turbulent flows. Therefore, Eq. (4.34) is also applicable for non–isotropic turbulence, provided the energy cascade is well established such that any anisotropy is lost by the time we reach the dissipative eddies.

4.6.2 Approximate Scaling of Production and Dissipation

Using the mean velocity and length scales of $\overline{U}$ and δ (see Fig. 4.4) and the turbulence scales of u, Λ, and λ, the terms in production and dissipation may be estimated as

$$-\overline{uv} \sim u^2 \tag{4.35}$$

$$\frac{\partial \overline{U}}{\partial y} \sim \frac{\overline{U}}{\delta} \tag{4.36}$$

$$\overline{\left(\frac{\partial u}{\partial x} \right)^2} \sim \frac{u^2}{\lambda^2} \tag{4.37}$$

Substituting Eqs (4.35 and 4.36) into Eq. (4.33), we have

$$P_{turb} \sim u^2 \frac{\overline{U}}{\delta} \qquad (4.38)$$

The order of magnitude expression

$$\varepsilon \sim \frac{\nu u^2}{\lambda^2} \qquad (4.39)$$

is obtained by substituting Eq. (4.37) into Eq. (4.34). For "local equilibrium" where $P_{turb} = \varepsilon$, the order of magnitude estimates yield

$$u^2 \frac{\overline{U}}{\delta} \sim \nu \frac{u^2}{\lambda^2} \qquad (4.40)$$

or

$$\frac{\delta}{\lambda} \sim \left(\frac{\overline{U}\delta}{\nu} \right)^{1/2} \qquad (4.41)$$

Note that λ, like η, becomes smaller as Re, the term in brackets on the right-hand side of Eq. (4.41), increases. This attests that a portion of the smallest scales of turbulence is truncated in scaled-down laboratory experiments and models. This could become a real challenge in situations where the small eddies play a nontrivial role in the problem under investigation.

Equation (4.41) is an order of magnitude expression relating dissipative scale λ with boundary-layer thickness δ in terms of a Reynolds number defined by the boundary-layer thickness. As such, it is not useful for flow turbulence other than boundary-layer flow over a smooth plate. We thus proceed to obtain an expression relating the flow turbulence parameters, including λ, only.

4.6.3 Relating Production and Dissipation Scales

Recall that one important characteristic of turbulence is that the smallest scales of motion that govern dissipation, $\varepsilon \sim \nu u^2 / \lambda^2$, always adjust themselves in size to accommodate changes in dissipation. Let us revisit the electric mixer example, but this time we assume that dissipation is chiefly associated with dissipating eddy size λ. If we increase the speed of the impeller by boosting the power input, the dissipation will also increase proportionally, that is

$$P_{turb} \sim \varepsilon \sim \nu u^2 / \lambda^2 \qquad (4.42)$$

In other words, augmenting P_{turb} results in an increase in u and/or a decrease in λ. Let the average size of the energy containing eddies be denoted by Λ. If the decay time of these energy containing eddies is 1.5 Λ/u, (the eddy travels roughly $\frac{1}{2}$ $\pi \Lambda$ distance, half a revolution, as it passes its energy down the cascade), then the energy transfer rate

$$\varepsilon \sim \text{kinetic energy per unit mass/decay time} \sim 3u^2/2/1.5\Lambda/u \quad (4.43)$$

or

$$\varepsilon \sim u^3 / \Lambda \qquad (4.44)$$

From the two estimates for dissipation, Eqs (4.42 and 4.44), we have

$$\varepsilon \sim \nu u^2 / \lambda^2 \sim u^3 / \Lambda \qquad (4.45)$$

This can be written as

$$\Lambda/\lambda \sim (u\Lambda / \nu)^{\frac{1}{2}} \qquad (4.46)$$

or

$$\Lambda/\lambda \sim Re_\Lambda^{\frac{1}{2}} \qquad (4.47)$$

We see that the difference between Λ and λ is less than that between Λ and η as described by Eq. (4.25). This indicates that λ is somewhat larger than η in the energy cascade. With increasing Reynolds number, λ decreases and even more so for η, resulting in reduction in η/Λ, λ/Λ, and η/λ. Recalling that Λ is more or less set by the physical dimensions involved, this implies that the energy cascade extends at the high-frequency, small scales end, especially right around the limit.

4.7 REFINED ESTIMATES FOR TURBULENCE DISSIPATION AND INTEGRAL SCALES

In the preceding sections, we have limited ourselves to "order of magnitude" estimates for the relationships between the different scales involved in a turbulent flow. Let us forge ahead to improve these approximations from "order of magnitude" to within "a factor or so."

4.7.1 Dissipation Microscales in Isotropic Turbulence

Improved estimates require a more precise definition of the turbulence dissipation scale λ, which so far has been used to approximate some typical scale that represents the range of eddy sizes that participate strongly in dissipation, while the Kolmogorov scale has been prescribed to signify the smallest eddy size. A more rigorous definition of the dissipative eddy size is the Taylor microscale formulated by Taylor (1935). Taylor defined a cross-stream scale λ_g and an along-stream scale λ_f

$$\lambda_g^2 \equiv \frac{2\overline{u^2}}{\overline{\left(\frac{\partial u}{\partial y}\right)^2}} \tag{4.48}$$

and

$$\lambda_f^2 \equiv \frac{2\overline{u^2}}{\overline{\left(\frac{\partial u}{\partial x}\right)^2}} \tag{4.49}$$

For isotropic turbulence with no preferential direction, Taylor related the x and y velocity derivatives through continuity to show that

$$\lambda_g = \lambda_f / \sqrt{2} \tag{4.50}$$

With the rigorously defined Taylor microscale, the dissipation expression for isotropic turbulence can be expressed as

$$\varepsilon = 15\nu \overline{\left(\frac{\partial u}{\partial x}\right)^2} = 15\nu \frac{2\overline{u^2}}{\lambda_f^2} = 30\nu \frac{\overline{u^2}}{\lambda_f^2} \tag{4.51}$$

This can be expressed in term of the cross-stream Taylor microscale,

$$\varepsilon = 15\nu \frac{\overline{u^2}}{\lambda_g^2} \tag{4.52}$$

The standard practice is to adopt the cross-stream scale λ_g as the typical microscale. It is interesting to compare this result with our earlier order of magnitude scale estimate which produced $\varepsilon \sim \nu\, u^2 / \lambda^2$. For isotropic turbulence $u^2 = \overline{u^2}$, and hence, we see that

$$\lambda_g = \sqrt{15}\,\lambda \tag{4.53}$$

Note that neither λ nor λ_g are true dissipation scales because they are both derived by assuming that the velocity scale of the dissipating eddies are the same as that of the large eddies, that is, both are equal to u. In a more general form, the dissipation scale relation may be expressed as

$$\varepsilon = v u_d^2 / \lambda_d^2 \tag{4.54}$$

where u_d and λ_d are activity scales specific to the range of dissipating eddies. It can be shown via spectral analysis of dissipation that the typical size of the most active dissipating eddies is normally (Hinze, 1975)

$$\lambda_d \sim 0.3 \lambda_g \tag{4.55}$$

Combining the above two dissipation expressions using this result gives

$$u_d \approx 1.2 \sqrt{\overline{u^2}} \tag{4.56}$$

This shows that $u_d = u$ is a fairly good assumption. In other words, while the size of the dissipative eddies are typically more than a couple of orders of magnitude smaller than the energy-containing eddies, their corresponding velocity scales are rather similar in magnitude. We recall from the example conveyed in Table 4.1 that this is also more or less true concerning the characteristic velocity associated with the Kolmogorov scale, where u_η is within an order of magnitude of u_Λ, even though η is about three orders of magnitude smaller than Λ.

Let us relax the dissipation expression to include non–isotropic turbulence, that is

$$\varepsilon \approx 5 v \frac{\overline{q^2}}{\lambda_g^2} \tag{4.57}$$

where the total velocity variance

$$\overline{q^2} = \overline{u^2} + \overline{v^2} + \overline{w^2} \tag{4.58}$$

Note that the approximate equality arises because we have invoked the isotropic dissipation assumption for non–isotropic turbulent flows. As such, Eq. (4.57) is, in a sense, exact for isotropic turbulence where $\overline{q^2} = 3\overline{u^2}$.

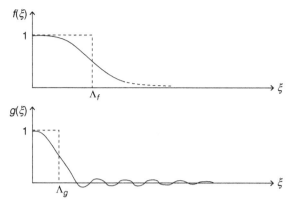

Figure 4.8 Along-stream and cross-stream auto-correlations. *(Created by B. Motameni).*

4.7.2 Integral Scales

Up to this point, Λ has been used to denote the size of some typical large eddy. We now proceed to refine the definition of the larger eddies in the energy cascade. We can follow Taylor (1935) and define the large length scales in terms of the cross-stream and along-stream space correlations. The along-stream correlation function is

$$f(\xi) = \frac{\overline{u(x,y)u(x+\xi,y)}}{\overline{u^2}} \tag{4.59}$$

and in the cross-stream direction

$$g(\xi) = \frac{\overline{u(x,y)u(x,y+\xi)}}{\overline{u^2}} \tag{4.60}$$

Here, u is the fluctuating velocity in the x direction, and ξ denotes a small displacement. These correlations are illustrated in Fig. 4.8. When ξ is equal to zero, the two streamwise fluctuations on the numerator are perfectly correlated, giving a correlation of unity. With increasing ξ, the correlation quickly decays and approaches zero. Figure 4.8 shows that the cross–stream correlation tends to oscillate around zero before it assumes zero with increasing separation ξ; the negative correlation occurs when two eddies revolve in opposing directions. This negative correlation is usually not observed in the streamwise correlation, presumably due to the prevailing convection of the mean velocity, which overshadows any small negative correlation. The integral scales are defined in terms of area under the correlations

$$\Lambda_f \equiv \int_0^\infty f(\xi)d\xi \tag{4.61}$$

and

$$\Lambda_g \equiv \int_0^\infty g(\xi)d\xi \tag{4.62}$$

Note that for any correlation function f and g that behave like $\exp(-\xi)$ as $\xi \to \infty$, it can be shown that there is a factor of 2, not $\sqrt{2}$, difference in scales defined by the two correlations; that is

$$\Lambda_g = \Lambda_f / 2 \tag{4.63}$$

in homogeneous isotropic turbulence.

With the well-defined integral length, we can tighten the meaning and definition of the somewhat loosely defined large eddy scale Λ. As the integral scale denotes the size of the energy-containing correlation length, it belongs to the lower-frequency, energy-producing end of the energy cascade; and so does Λ. Therefore, these two length scales are expected to be linearly related, that is

$$\Lambda = C \Lambda_g \tag{4.64}$$

where C is a constant. According to Hinze (1975)

$$\Lambda = 2.66 \Lambda_g = 1.33 \Lambda_f \tag{4.65}$$

We may interpret this as: Λ signifies the largest eddies, Λ_g the integral length, and both are engaged with turbulent kinetic energy production, supplying energy down the smaller scales along the energy cascade.

With the key scales more rigorously defined, the notion of the turbulence energy cascade can be improved as shown in Fig. 4.9. We note that the energy cascade is bounded by the largest eddies approximately 2.66 times the cross-stream integral length at the low-frequency end, and the smallest Kolmogorov scale at the high-frequency limit. Even though the smallest Kolmogorov scale may be most effective in dissipating kinetic energy into heat, most dissipation actually takes place via λ_d, which is roughly five times η, or about one-third λ_g.

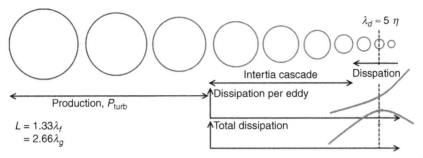

Figure 4.9 Refined turbulence energy cascade from production to dissipation. *(Created by D. Ting).*

4.8 TURBULENT KINETIC ENERGY SPECTRUM

The turbulent kinetic energy associated with the cascade of the large-to-small spectrum of eddies can be viewed in terms of the spectral distribution of energy. The Fourier decomposition can be expressed in terms of number of cycles per unit length, i.e., **wavenumber**, κ_w, or its inverse, **wavelength**, λ_w. Note that the wavelength is akin to length scale, that is, the size of the eddy, and the wavenumber is similar to the frequency. The large scales have longer wavelengths and lower frequencies, while the small scales have shorter wavelengths and higher frequencies. The turbulence kinetic energy per unit mass is thus

$$ke = \int_0^\infty E\left(\kappa_w\right) d\kappa_w \tag{4.66}$$

where $E(\kappa_w)d\kappa_w$ is the amount of kinetic energy possessed by the eddies with wavenumbers between κ_w and $\kappa_w + d\kappa_w$. The **energy spectral density** or **energy spectrum function** $E(\kappa_w)$ is a function of the energy-containing large eddies Λ and the mean strain rate which transfers the energy from the mean flow to the large eddies. The amount of energy $E(\kappa_w)$ is also dependent on the rate of dissipation, that is, ε and ν. For well-developed turbulence where Λ is much larger than η, we have

$$\varepsilon \sim ke^{3/2}/\Lambda \tag{4.67}$$

as conjured by dimensional analysis, and confirmed by measurements (Taylor, 1935). In other words

$$ke \sim (\varepsilon\Lambda)^{2/3} \tag{4.68}$$

According to the energy cascade (Fig. 4.10), for well-developed turbulence, there is a range of wavenumbers over which neither production nor

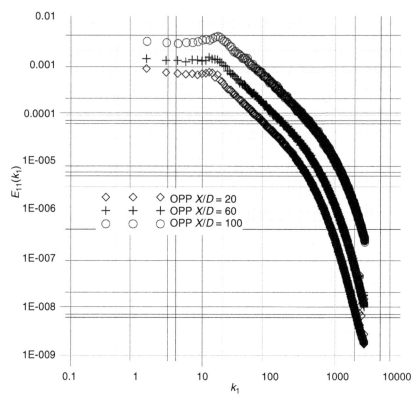

Figure 4.10 Orificed, perforated plate-generated streamwise turbulence velocity spectra at $U = 10.8$ m/s. *(Created by R. Liu).*

dissipation is dominant; that is, the inertial transfer of kinetic energy is the prominent player. Over this **inertial sub-range**, $E(\kappa_w)$ depends only on ε and κ_w. Based on dimensional analysis, Kolmogorov came to the conclusion that

$$E(\kappa_w) = C_\kappa\, \varepsilon^{2/3}\, \kappa_w^{-5/3} \tag{4.69}$$

where $1/\Lambda << \kappa_w << 1/\eta$, and C_κ is the Kolmogorov constant.

Figure 4.10 shows some typical streamwise turbulence velocity spectra downstream of an orificed, perforated plate turbulence generator at 10.8 m/s wind (Liu et al., 2007); we will expound on this unique, passive turbulence generation in the ensuing chapter on grid turbulence. Accordingly, more will be said about the energy spectrum in general. Briefly, k is the frequency, which is inversely proportional to the size of the eddying motion. It is clear that most of the energy is associated with the low-frequency, large eddies, and the amount of energy drops sharply with increasing k. Most interestingly, we see the separation of the highest-frequency, dissipating, viscous range from the lowest-frequency production range by the inertial sub-range with a $-5/3$ slope in Fig. 4.10. We will discuss this further when looking at grid turbulence.

Problems

Problem 4.1 Turbulence from volcanic eruption
A turbulent plume is generated via a volcanic eruption where the integral scale $\Lambda \sim 20$ m and a typical turbulent velocity is 30 m/s. If the viscosity of the volcanic gas is 10^{-5} m²/s, estimate λ and η. How do these small scales compare to the mean free path length?

Problem 4.2 Molecular versus turbulent diffusion
A laboratory is roughly 4 m × 4 m × 4 m. Near the center of a wall is a floor heater. Assume that the air just above the heater is 20°C above the room temperature.
1. How long does it take to heat up the laboratory if we assume the air inside the lab is stagnant?
2. From the convection of the heated air plumes, estimate the turbulent viscosity. How long does it take to heat up the lab when we include the turbulent effect?

Problem 4.3 Isotropic turbulence in a spherical container
A 10 cm radius sphere containing water at standard temperature and pressure is being stirred at a rate where the power input into the volume of water is 500 W. At equilibrium, roughly isotropic turbulence (outside of the boundary layer) is maintained.
1. What would happen to the turbulence intensity u_{rms} and the sizes of the large and small eddies if the stirring intensity is doubled, i.e., doubling the kinetic energy input rate?
2. What would happen to the turbulence intensity and the sizes of the large and small eddies if the size (radius) of the sphere is doubled while keeping the same stirring intensity?
3. What would happen to the turbulence intensity and the sizes of the large and small eddies if the fluid viscosity is doubled (for the same sphere size and stirring intensity)?
4. Describe how the turbulence intensity and the sizes of the large and small eddies vary if the stirring is suddenly stopped (from equilibrium).
5. Describe how the turbulence intensity and the sizes of the large and small eddies vary when the stirring is initiated from an initially stagnant volume of water.

Problem 4.4 Turbulent spreading

A certain amount of hot fluid is released in a turbulent flow with characteristic velocity U and characteristic length l. The temperature of the patch is somewhat ($\sim 10°C$) higher than the ambient temperature, but the density difference and the effect of buoyancy may be neglected. Estimate the rate of spreading of the patch and the rate at which the maximum temperature difference decreases. Assume that the size of the patch at the time of release is much smaller than l and much larger than Kolmogorov microscale η. The use of eddy diffusivity is appropriate, but be careful when choosing the velocity and length scales that are needed to form an eddy diffusivity, in particular, as long as the size of the patch remains smaller than the length scale l.

REFERENCES

Anderson, Jr, J.D., 2005. Ludwig Prandtl's boundary layer, Physics Today: 42–48.

Gad-el-Hak, M., 1998. Fluid mechanics from the beginning to the third millennium. Int. J. Eng. Edu. 14 (3), 177–185.

Hinze, J.O., 1975. Turbulence, second ed. McGraw-Hill, USA.

Kolmogorov, A.N., 1941. The local structure of turbulence in incompressible viscous fluid for very large Reynolds numbers. Dokl. Akad. Nauk SSSR 30, 299–303.

Liepmann, H.W., 1979. The rise and fall of ideas in turbulence. Am. Sci. 67 (2), 221–228.

Lim, T.T., 1989. A vortex ring interacting with an inclined wall. Exp. Fluids 7, 453–463.

Lim, T.T., Nickels, T.B., 1992. Instability and reconnection in the head-on collision of two vortex rings. Nature 357, 225–227.

Liu, R., Ting, D.S-K., Checkel, M.D., 2007. Constant Reynolds number turbulence downstream of an orificed perforated plate. Exp. Therm. Fluid Sci. 31, 897–908.

Prandtl, L., 1904. Über flüssigkeitsbewegung bei sehr kleiner reibung, Verhandlungen des dritten internationalen Mathematiker-Kongresses, Heidelberg, Germany, pp. 484–491. Reprinted 2006 by Cambridge University Press with a new introduction by Peter Lynch.

Richardson, L.F., 1922. Weather Prediction by Numerical Process. Cambridge University Press, Cambridge.

Taylor, G.I., 1935. Statistical theory of turbulence. Proc. R. Soc. Lond. Ser. A 151, 421–444.

Tennekes, H., Lumley, J.L., 1972. A First Course in Turbulence. MIT Press, Cambridge.

Vassilicos, J.C., 2015. Dissipation in turbulent flows. Ann. Rev. Fluid Mech. 47, 95–114.

Wilson, D.J., 1989. Mec E 632 Turbulent Fluid Dynamics. University of Alberta, Edmonton.

CHAPTER 5

Turbulence Simulations and Modeling

It is far better to foresee even without certainty than not to foresee at all.

– Henri Poincaré

Contents

Chapter Objectives

- To understand what turbulence modeling is and why it is needed.

- To discern and differentiate the two extremes of turbulence modeling, from directly solving the Navier-Stokes equations to modeling all scales of turbulence.

- To learn about the "closure problems" in turbulent flows.

- To have a conceptual understanding of the zero-order closure, one-equation, and two-equation models.

- To appreciate the notions of large eddy simulation and direct numerical simulation.

NOMENCLATURE

C	Constant, coefficient
d	Dimensional
DES	Detached eddy simulation
DNS	Direct numerical simulation
F	Force
h	Numerical resolution
k	Turbulence kinetic energy per unit mass of fluid specific turbulence kinetic energy
l	Length, large turbulence length scale
l_{mfp}	Mean free path
l_{mix}	Mixing length
LES	Large eddy simulation

mix	Mixing
p	The fluctuating component of the pressure
P	Time-averaged pressure
RANS	Reynolds-averaged Navier-Stokes
Re	Reynolds number
RNG	Re-normalized group
rms	Root mean square
S	Stress (tensor)
t	Time
turb	Turbulent
u	The fluctuating component of the velocity (in the x direction)
U	Time-averaged velocity (in the x direction)
v	The fluctuating component of the velocity in the y direction
v_{mix}	Mixing velocity
v_{th}	Thermal (average molecular) velocity
V	Time-averaged velocity in the y direction
w	The fluctuating component of the velocity in the z direction
W	Time-averaged velocity in the z direction
x	Distance in the x (streamwise) coordinate
y	Distance in the y (cross-stream, vertically up) coordinate
z	Distance in the z (cross-stream, out of the page) coordinate

Greek Symbols

α	Eddy acceleration
Δ_{LES}	Filter length for LES
δ_{ij}	Kronecker delta, $\delta_{ij} = 1$ if $i = j$, $\delta_{ij} = 0$, if $i \neq j$
η	Kolmogorov length scale
κ	Von Karman constant
κ_η	Kolmogorov wavenumber
λ	Taylor microscale
μ	Dynamic viscosity
μ_{turb}	Turbulence dynamic viscosity
v	Kinematic viscosity, $v = \mu/\rho$
v_{turb}	Turbulence kinematic viscosity
ρ	Density
τ	Shear
ω	Specific dissipation rate
ε	Dissipate rate

5.1 INTRODUCTION

The renowned Navier-Stokes equations can describe both laminar and turbulent flows. It is, however, neither practical nor feasible, at present and possibly also into the future, to solve the nonlinear equations of motion for instantaneous velocities of engineering turbulence problems. We need approximate and semi-empirical methods to predict the effect of flow turbulence in engineering applications. Also noteworthy is that in practice, the details associated with an exact solution are usually unnecessary. In other words, an approximate solution, which captures the essence of the problem is often all that is needed.

Generally speaking, turbulence modeling is simply a means of describing the phenomenon of turbulent flow somewhat quantitatively over a range of conditions. This implies that turbulence modeling is a procedure that can deduce preponderant and palpable turbulent variables without directly solving the time-dependent Navier-Stokes equations. Specifically, functional relationships between the Reynolds stresses and the mean-flow characteristics are hypothesized, and "standard experiments" are conducted to provide values for the constants associated with these functional relationships. A good model is typically one which introduces the minimum amount of complexity while capturing the essential underlying physics.

There are two extremes when it comes to turbulence modeling. The familiar models based on the Reynolds-averaged Navier-Stokes (RANS) equations address all scales of eddies, from the largest to the smallest; that is, none of the eddies are deduced directly from the Navier-Stokes equations. At the other extreme is direct numerical simulation (DNS), where all turbulent eddies are explicitly resolved; that is, the flow characteristics manifested by all eddies are calculated directly via the equations of motion. As such, strictly speaking, DNS is not considered modeling in the context of the resolving of Reynolds stresses. With increasing Reynolds numbers, the resolving of a growing number of progressively smaller momentaneous eddies forbids the practicality of DNS. For this reason, large eddy simulation (LES) and its counterparts which resolve eddies (but only the large ones) emerge as a versatile compromise.

Only a brief overview of the classical and standard models is included here for the sake of completeness. Monographs, which have specialized in turbulence modeling, include Wilcox (1993), Chen and Jaw (1997), Pope (2000), and Wilcox (1998, 2006). The purpose of this chapter is to provide nonexperts with the stepping stool from which they can easily step up into these specialized treatises.

Recall from Chapter 2 that for incompressible, isothermal, laminar flows, the momentum equation can be expressed as

$$\rho \frac{DU_i}{Dt} = \rho F_i - \frac{\partial P}{\partial x_i} + \mu \frac{\partial^2 U_i}{\partial x_j \partial x_j} \qquad (5.1)$$

or simply

$$\rho \frac{DU_i}{Dt} = \rho F_i + \frac{\partial \tau_{ji}}{\partial x_j} \qquad (5.2)$$

The Newtonian stress-rate of strain relationship is

$$\tau_{ij} = -P\delta_{ij} + 2\mu \left(S_{ij} - \frac{1}{3}\delta_{ij} \frac{\partial U_k}{\partial x_k} \right) \qquad (5.3)$$

where μ is viscosity and S_{ij} is the mean rate of strain tensor

$$S_{ij} = \frac{1}{2}\left(\frac{\partial U_i}{\partial x_j} + \frac{\partial U_j}{\partial x_i} \right) \qquad (5.4)$$

δ_{ij} is the Kronecker delta; that is, $\delta_{ij} = 1$ if $i = j$ and $\delta_{ij} = 0$ otherwise.

For fluctuating turbulent flows, the stream-wise, x component of the momentum equations for the incompressible and isothermal case can be expressed as

$$\frac{\partial \tilde{U}}{\partial t} + \frac{\tilde{U}}{}\frac{\partial \tilde{U}}{\partial x} + \frac{\tilde{V}}{}\frac{\partial \tilde{U}}{\partial y} + \frac{\tilde{W}}{}\frac{\partial \tilde{U}}{\partial z} = -\frac{(1/\rho)\,\partial \tilde{P}}{\partial x} + v\left(\frac{\partial^2 \tilde{U}}{\partial x^2} + \frac{\partial^2 \tilde{U}}{\partial y^2} + \frac{\partial^2 \tilde{U}}{\partial z^2} \right) \qquad (5.5)$$

where $\tilde{U} = U + u$, $\tilde{V} = V + v$, $\tilde{W} = W + w$, and $\tilde{P} = P + p$. Recalling that the instantaneous velocity, $\tilde{U}$ = time-averaged or mean velocity U + the fluctuating component u; that is, $u = \tilde{U} - U$. As detailed in Chapter 2, we can apply Reynolds decomposition and time averaging to Eq. (5.5). After some manipulations, we get, for the x component

$$\rho\left(U\frac{\partial U}{\partial x} + V\frac{\partial U}{\partial y} + W\frac{\partial U}{\partial z} \right) = -\frac{\partial P}{\partial x} + \frac{\partial}{\partial x}\left(\mu\frac{\partial U}{\partial x} - \rho\overline{u^2} \right)$$

$$+ \frac{\partial}{\partial y}\left(\mu\frac{\partial U}{\partial y} - \rho\overline{uv} \right) + \frac{\partial}{\partial z}\left(\mu\frac{\partial U}{\partial z} - \rho\overline{uw} \right) \qquad (5.6)$$

where the second term on the right-hand side is the normal stress, the third term is the shear on the x-z plane, and the last term is the shear on the x-y plane. In short, for the ith component

$$U_j \frac{\partial U_i}{\partial x_j} = -\frac{1}{\rho}\frac{\partial P}{\partial x_i} + v\frac{\partial^2 U_i}{\partial x_j \partial x_j} - \frac{\partial \overline{u_i u_j}}{\partial x_j} \qquad (5.7)$$

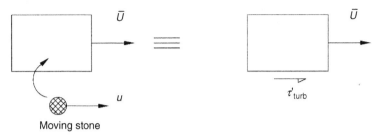

Figure 5.1 Reynolds shear stresses – transport of a moving stone into a smoothly moving wagon. *(Created by N. Cao).*

or

$$U_j \frac{\partial U_i}{\partial x_j} = -\frac{1}{\rho}\frac{\partial P}{\partial x_i} + \frac{1}{\rho}\frac{\partial}{\partial x_j}\left(\mu\frac{\partial U_i}{\partial x_j} - \rho\overline{u_i u_j}\right) \qquad (5.8)$$

We note that the term $\dfrac{\partial}{\partial x_j}\left(\overline{u_i u_j}\right)$ is the mean transport of fluctuating momentum by the turbulent velocity fluctuations. It exchanges momentum between the turbulence and the mean flow, even though the mean momentum $\rho\overline{u_i}$ of the turbulent fluctuations is zero. Physically we may visualize turbulent transport as follows: The diagonal terms $\rho\overline{u^2}$, $\rho\overline{v^2}$, and $\rho\overline{w^2}$ represent fluctuating normal stresses, and their contribution to the mean momentum transport are typically secondary in comparison to the cross-correlations. The cross-correlations, such as $\overline{uv}, \overline{vw}$, signify the turbulent shear stresses, and they can drastically enhance the transport phenomenon. As discussed in Chapter 2, these shear stresses may be pictured as Zorro moving at a different speed than a moving wagon, or as a stone being thrown into a moving wagon (see Fig. 5.1). In other words, the forward or backward jerk of an otherwise steady wagon caused by a piece of stone represents the shear stress induced by turbulence on the otherwise smooth laminar flow.

5.2 THE CLOSURE PROBLEM IN TURBULENT FLOWS

In the Cartesian coordinate system, the basic system of four equations are: (1) continuity, (2) x momentum, (3) y momentum, and (4) z momentum. The four variables which need to be solved for in laminar flow are P, U, V, and W, where P is pressure, U is streamwise velocity, V is the velocity in the

y direction, and W is the velocity in the z direction. The turbulent momentum equations contain the terms of the Reynolds stress tensor

$$\vec{\tau} = -\rho \begin{bmatrix} \overline{u^2} & \overline{uv} & \overline{uw} \\ \overline{vu} & \overline{v^2} & \overline{vw} \\ \overline{wu} & \overline{wv} & \overline{w^2} \end{bmatrix} \tag{5.9}$$

As $\overline{uv} = \overline{vu}$, $\overline{uw} = \overline{wu}$ and $\overline{vw} = \overline{wv}$, there are only six independent variables. In short, we have ten variables (three velocity components, one pressure, and six Reynolds stress terms), but only four equations (one continuity and three momentum); that is, we need six additional equations to solve the six new unknowns. We can try to generate extra equations by taking moments of the Navier-Stokes equations, but this will lead to extra variables. The so-called "closure problem" is the challenge associated with finding supplementary relationships for the unknown correlations. These relationships can be algebraic expressions (for zero-order closures) or additional differential equations. We define the "order of closure" (n-equation model) as the number of differential transport equations required in addition to those expressing conservation of mass, momentum, and energy.

5.2.1 Zero-Order Closures (Algebraic Models)

The simplest of all turbulence closures are strictly algebraic. As such, they apply only to the "simplest" of turbulent flows. The following is an abbreviated historic account of this.

Boussinesq (1877, 1897): In 1877, Boussinesq introduced the eddy-viscosity approximation, thereby allowing one to approximate the turbulent flow by assigning a quantitative value to the eddy (dynamic) viscosity. He modeled the turbulent stresses responsible for significantly augmenting the molecular counterpart within this eddy viscosity. In the rudimentary form, this eddy viscosity is assigned a fixed value estimated from limited experiments. Hence, we have a uniform eddy viscosity throughout the flow field under consideration; nonetheless, we could divide a flow field into regions of various uniform eddy viscosities, as seen in wall-bounded flows. As simple as it is, Boussinesq's eddy-viscosity model has been shown to work well in a limited sense for some "calibrated" free-shear flows such as axisymmetric jets, 2D jets, and mixing layers.

Specifically, Boussinesq suggested that a turbulent flow could be regarded as having an enhanced dynamic viscosity, also called a turbulent or eddy viscosity μ_{turb}. The turbulent shear stress

$$\tau_{xy} = -\rho\overline{uv} = \mu_{turb}\frac{\partial U}{\partial y} \tag{5.10}$$

Recall that the shear associated with the laminar counterpart is simply

$$\tau_{xy} = \mu\frac{dU}{dy} \tag{5.11}$$

As highlighted by Garde (2010), μ_{turb} is generally much larger than the fluid dynamic viscosity μ. Note that while μ is a constant for a given fluid at a specified state, μ_{turb} is a function of both flow condition and fluid density. In other words, the fluid viscosity is a property of the fluid, which is specified by the thermodynamic state that the fluid is in, whereas the eddy viscosity depends also on the specific flow conditions involved. Hence, standard flow experiments need to be carried out to quantify μ_{turb} before we can apply it to problems with similar conditions.

The Boussinesq hypothesis can be somewhat generalized by using the kinematic viscosity as opposed to the dynamic viscosity. The resulting general expression is

$$\tau_{ij} - \frac{1}{3}\tau_{kk}\delta_{ij} = \rho v_{turb}\left(\frac{\partial U_i}{\partial x_j} + \frac{\partial U_j}{\partial x_i}\right) \tag{5.12}$$

Boussinesq assumed that the turbulent or eddy kinematic v_{turb}, the only empirical parameter, is a constant. For simple free-shear flows, the eddy viscosity v_{turb} has been found to be roughly constant at one to two orders of magnitude above the laminar value. For wall-bounded flows, on the other hand, v_{turb} varies significantly depending on its position, starting with zero at the wall, as expected.

Prandtl (1925) furthered Boussinesq's eddy-viscosity concept to include the mixing-length notion, along with the concept of a boundary layer. In his attempt to express eddy viscosity in terms of flow conditions, Prandtl introduced mixing length. This mixing length is analogous to the mean free path of a gas as deduced via the kinetic theory. According to this theory, the molecular viscosity of the fluid at a given state μ is equal to $\frac{1}{2}\rho v_{th} l_{mfp}$, where v_{th} is the thermal velocity (average molecular velocity) and l_{mfp} is the mean free path. Note that the actual value of viscosity is $\mu = 0.499\,\rho\,v_{th}\,l_{mfp}$ (Jeans, 1962). Hence, for laminar flows, we have

$$\tau_{xy} = \tfrac{1}{2}\rho v_{th} l_{mfp} \frac{dU}{dy} \tag{5.13}$$

The corresponding expression for turbulent flows is thus

$$\tau_{xy} = \frac{1}{2}\rho v_{mix} l_{mix} \frac{dU}{dy} \tag{5.14}$$

where v_{mix} is the mixing velocity and l_{mix} is the mixing length. As a first approximation, the mixing velocity can simply be equated to the fluctuating turbulent velocity. As such, this mixing-length model is an algebraic model or a zero-equation model, where no additional equation is required. It is in contrast to n-equation model, which is a model that requires solution of n additional differential transport equations other than those expressing conservation of mass, momentum, and energy.

In short, Prandtl (1925) accounted for the variability of turbulent mixing with only one empirical constant: the mixing length. This mixing length intends to represent a distance within which fluid particles coalesce into lumps that cling together and move as a unit; that is, the length scale encloses a lump of fluid swirling around in some cohesive manner. For typical boundary layer flows, the mixing length is the lateral (perpendicular to the boundary) length. In other words, a fluid element displaced vertically from its original position y in the boundary layer would retain more or less its original streamwise velocity, $U(y)$, at that level y. The apparent perturbation velocity for an element displaced vertically at a small distance l_{mix} is

$$u = U(y) - U(y + l_{mix}) - l_{mix}\frac{\partial U}{\partial y} \tag{5.15}$$

One major physical assumption behind this is that streamwise pressure forces and viscous stresses are unimportant. This is relatively valid for three-dimensional eddies that are "flat," that is, have horizontal dimensions much larger than the vertical ones (Russel and Landahl, 1984). We can thus relate the mixing velocity with the mixing length and the velocity gradient

$$v_{mix} = \text{constant} \cdot l_{mix} \cdot |\,dU/dy\,| \tag{5.16}$$

It should be noted that this implies that the mixing velocity is locally resolved from the mean velocity gradient (Pope, 2000). In particular, it predicts a zero-mixing velocity when the gradient is zero when, in fact, there are many situations in which this is not true. Nonetheless, with Eq. (5.16), we can rewrite the turbulent shear as

$$\tau_{xy} = \frac{1}{2}\rho d^2_{mix}\left|\frac{dU}{dy}\right|\frac{dU}{dy} \tag{5.17}$$

We see that the eddy viscosity can thus be expressed as

$$\mu_{turb} = \rho l^2_{mix}\left|\frac{dU}{dy}\right| \tag{5.18}$$

The corresponding kinematic eddy viscosity is

$$\nu_{turb} = l^2_{mix}\left|\frac{dU}{dy}\right| \tag{5.19}$$

Thus, we have expressed the Reynolds shear stress in terms of the mixing length

$$-\overline{uv} = \left|l_{mix}\frac{dU}{dy}\right|^2 \tag{5.20}$$

In turbulent shear flows, $\left|-\overline{uv}\right| \approx 0.4 u_{rms} v_{rms}$ (Townsend, 1976), where 0.4 is the von Karman constant. For a boundary layer, we may assume $l_{mix} \propto y$, the distance from the wall; that is, in wall-bounded flows, the van Driest (1956) model assumes

$$l_{mix} = ky\left[1 - \exp\left(\frac{y^+}{A_0^+}\right)\right] \tag{5.21}$$

where $A_0^+ = 26$, $y^+ = y\,u^*/\nu$, $u^* = \sqrt{(\tau_w/\rho)}$, τ_w is the wall shear stress (Smith and Cebeci, 1967; Baldwin and Lomax, 1978).

We note that the major drawback of Prandtl's mixing-length theory is that the mixing length l_{mix}, which must be known in order to solve the problem, is different for each flow. In spite of its theoretical shortcomings, the mixing-length model does an excellent job of reproducing measurements. Consequently, eddy-viscosity models based on mixing-length theory have been fine-tuned for many flows; see Cebeci and Smith (1974), among many others. The mixing-length computation has been found to be quite accurate for "predicting" equilibrium turbulent flows, in which the turbulent properties vary very slowly. For more recent (post-1950) advancement of zero-equation models, see van Driest (1956), Cebeci and Smith (1974), Baldwin and Lomax (1978), among others. For example, Smagorinsky (1963) proposed

$$v_{\text{turb}} = \Delta x \Delta y \sqrt{\left(\frac{\partial U}{\partial x}\right)^2 + \left(\frac{\partial V}{\partial y}\right)^2 + \frac{1}{2}\left(\frac{\partial U}{\partial y} + \frac{\partial V}{\partial x}\right)^2} \tag{5.22}$$

for numerical models. Most interestingly, this elementary zero-equation approach still finds its use in some applications today; see Ng et al. (2011), Bazargan and Mohseni (2012), Li et al. (2012), and Alammar (2014).

5.2.2 One-Equation Models

Both Kolmogorov (1942) and Prandtl (1945) recognized that the weakness of the mixing-length model was its need to choose the velocity scale via $l_{\text{mix}} \, \partial U / \partial y$, which requires both empirical deduction of l_{mix} and the foreknowledge of the $\partial U / \partial y$ involved. Twenty years after his original mixing-length model, Prandtl introduced the eddy-viscosity model in 1945; that is, he expressed the eddy viscosity as a function of the turbulent kinetic energy per unit mass, k, directly. Specifically, Prandtl (1945) postulated that instead of setting $v_{\text{mix}} \approx l_{\text{mix}} \, |dU/dy|$, we can compute v_{mix} "directly." Hence, the birth of the very first one-equation model, where a single transport equation for turbulent viscosity is solved. The actual development is rather involved and of varied methods. We will simply highlight the concept behind the one-equation approach below.

In the Cartesian coordinate system, we have contribution to turbulent kinetic energy from the x, y, and z directions. The turbulent kinetic energy per unit mass, the specific turbulence kinetic energy

$$k = \frac{1}{2}\left(\overline{u^2} + \overline{v^2} + \overline{w^2}\right) \tag{5.23}$$

Hence, we see that k is related to the trace of the Reynolds-stress tensor as

$$k = \frac{1}{2}\overline{u_i^2} = \frac{1}{2}\tau_{ii} \tag{5.24}$$

Via dimensional analysis (Wilcox, 2006), we can relate this to the kinematic eddy viscosity by

$$\nu_{turb} = \text{constant } k^{1/2} \, l \tag{5.25}$$

where the characteristic length l signifies a turbulence length scale similar to the mixing length. Note that we may follow Pope (2000) and simply take the mixing length as the turbulent length scale, that is, use l and l_{mix} interchangeably.

First, Kolmogorov (1942), and subsequently but independently, Prandtl (1945), proposed a model transport equation for the specific turbulent kinetic energy k, from which k may be deduced. This is hence the first one-equation model, as it is used to resolve just one turbulence quantity, that is, the specific turbulent kinetic energy k. Following Pope (2000), the model transport equation for k is

$$\frac{Dk}{Dt} + \nabla \cdot \left[\frac{1}{2}\overline{u_i u_i u_j} + \frac{\overline{u_i \tilde{p}}}{\rho} - 2\nu \overline{u_j \frac{1}{2}\left(\frac{\partial u_i}{\partial x_j} + \frac{\partial u_j}{\partial x_i}\right)} \right] = -\overline{u_i u_j}\frac{\partial U_i}{\partial x_j} - \varepsilon \tag{5.26}$$

We note that the first term on the right-hand side is the turbulence production term. This and the total change in the specific turbulent kinetic energy are in "closed form," that is, they are explicitly expressed in terms of known variables. The other terms, however, are "open" in the sense that they need "closure approximations" that relate the unknowns in terms of known variables. Consequently, the following are the essential components associated with the one-equation model (Pope, 2000):

1. The mixing length, l_{mix}, which needs to be specified;
2. The specific turbulent kinetic energy, k, deduced from the equation, that is, Eq. (5.26);
3. The turbulent or eddy viscosity, $\nu_{turb} = \text{constant } k^{1/2} l_{mix}$;
4. The Reynolds stresses

$$\overline{u_i u_j} = \frac{2}{3}k\delta_{ij} - \nu_{turb}\left(\frac{\partial U_i}{\partial x_j} + \frac{\partial U_j}{\partial x_i}\right) \tag{5.27}$$

5. The velocity and pressure fields determined from the Reynolds equations.

Other than a few noted exceptions, one-equation models have proven to be relatively unsuccessful as they are incomplete in the sense that only k is described mathematically. Notwithstanding this drawback, they provide a step to more successful, higher-order models.

5.2.3 Two-Equation Models

The basis of two-equation modeling is to find or derive a transport equation for the length scale l or another appropriate quantity. In addition to the (transport) equation describing the turbulence kinetic energy per unit mass, k (and hence, u), an equation expressing the characteristic turbulent length scale l, or another appropriate quantity, is also introduced. Therefore, two-equation models are "complete" in the sense that they can be used to "predict" turbulent properties of a flow without prior knowledge of the turbulence structure. The two acclaimed two-equation models are the k-ω and the k-ε models. A brief description of each is provided here.

5.2.3.1 The k-ω model
According to Boussinesq (1877)

$$\tau_{ij} = 2\nu_{turb}S_{ij} - \frac{2}{3}k\delta_{ij} \qquad (5.28)$$

This is referred to as the Boussinesq hypothesis, the Boussinesq eddy-viscosity assumption, or simply the Boussinesq approximation. According to this equation, ν_{turb} is plausibly proportional to k (Wilcox, 2006). Furthermore, from dimensional analyses, we have

$$\nu_{turb} \sim k / \omega \qquad (5.29)$$

$$l \sim k^{1/2} / \omega \qquad (5.30)$$

$$\varepsilon \sim \omega k \qquad (5.31)$$

We note that $1/\omega$ is a turbulence time scale, and, as per Eq. (5.31), it is the time scale associated with turbulent kinetic energy dissipation. The turbulent length scale, which is analogous to the mixing length is expressed by $\sqrt{k}/\omega$. Therefore, ω may be an appropriate quantity which can be used to substitute the mixing length. Consequently, the two equations are the specific turbulent kinetic energy equation, which can be expressed in the form

$$\frac{\partial k}{\partial t} + U_j \frac{\partial k}{\partial x_j} = \tau_{ij} \frac{\partial U_i}{\partial x_j} - \varepsilon + \frac{\partial}{\partial x_j}\left[\left(\nu + \frac{\nu_{turb}}{C_k}\right)\frac{\partial k}{\partial x_j}\right] \qquad (5.32)$$

and an equation for ω, which could be

$$\frac{\partial \omega}{\partial t} + U_j \frac{\partial \omega}{\partial x_j} = -C_{\omega 2}\omega^2 + \frac{\partial}{\partial x_j}\left(C_\omega \nu_{turb}\frac{\partial \omega}{\partial x_j}\right) \qquad (5.33)$$

as noted by Kolmogorov (1942) and Wilcox (2006). Keep in mind that C_k, $C_{\omega 2}$ and C_ω are closure coefficients which need to be deduced from standard experiments.

The earlier versions of this first "complete" model of turbulence appeared to have many challenges. Wilcox (2006) seems to have overcome some of these earlier hurdles, significantly improving the performance of k-ω model.

5.2.3.2 The k-ε model
In 1945, Chou proposed modeling the exact equation for the specific turbulent dissipation rate ε, which led to

$$\nu_{turb} \sim k^2 / \varepsilon \qquad (5.34)$$

$$l \sim k^{3/2} / \varepsilon \qquad (5.35)$$

It is clear that the two equations from this approach are the k equation and the ε equation, and hence, the k-ε model. After Chou (1945), further advancement along this line was made by Davydov (1961), as well as Harlow and Nakayama (1968), in particular. Noteworthily, Jones and Launder (1972) and Launder and Sharma (1974), among others, led to the popularity of this k-ε model.

Briefly, beginning with Eqs (5.28 and 5.32), the idea is to derive the exact equation for ε. This can be done by taking the following moment of the Navier-Stokes equation

$$\overline{2\nu \frac{\partial u}{\partial x_j} \frac{\partial}{\partial x_j}\left(\rho \frac{\partial \tilde{U}_i}{\partial t} + \rho \tilde{U}_k \frac{\partial \tilde{U}_i}{\partial x_k} + \frac{\partial \tilde{P}}{\partial x_i} - \mu \frac{\partial^2 \tilde{U}_i}{\partial x_k \partial x_k}\right)} = 0 \qquad (5.36)$$

where the term in the bracket is called the "Navier-Stokes operator." After some amount of manipulations, we get

$$
\frac{\partial \varepsilon}{\partial t} + U_j \frac{\partial \varepsilon}{\partial x_j} = \left[-2v \left(\overline{\frac{\partial u_i}{\partial x_k} \frac{\partial u_j}{\partial x_k}} + \overline{\frac{\partial u_k}{\partial x_i} \frac{\partial u_k}{\partial x_j}} \right) \frac{\partial U_i}{\partial x_j} - 2v \overline{u_k \frac{\partial u_i}{\partial x_j}} \frac{\partial^2 U_i}{\partial x_k \partial x_j} \right]
$$

$$
- \left[2v \overline{\frac{\partial u_i}{\partial x_k} \frac{\partial u_i}{\partial x_m} \frac{\partial u_k}{\partial x_m}} + 2v^2 \overline{\frac{\partial^2 u_i}{\partial x_k \partial x_m} \frac{\partial^2 u_i}{\partial x_k \partial x_m}} \right] \quad (5.37)
$$

$$
+ \frac{\partial}{\partial x_j} \left[v \frac{\partial \varepsilon}{\partial x_j} - v \overline{u_j \frac{\partial u_i}{\partial x_m} \frac{\partial u_i}{\partial x_m}} - 2 \frac{v}{\rho} \overline{\frac{\partial p}{\partial x_m} \frac{\partial u_j}{\partial x_m}} \right]
$$

where the terms in square brackets [ ] signify rate of production of dissipation, rate of destruction of dissipation, the sum of molecular diffusion of dissipation, and turbulent transport of dissipation.

There are a colossal number of publications on k-ω and k-ε models. In addition to these two popular models, there are also many other two-equation models, as well as other multi-equation models; see Table 6.2 of Garde (2010), which is compiled from Launder and Spalding (1972). Interested readers are referred to the specialized monographs on turbulent modeling prior to diving into the ocean of journal publications.

It is worth stressing that the major assumption embodied in the closures so far is that of the Prandtl-Kolmogorov-Bousinesq assumption, which begins with the gradient diffusion hypothesis applied to the stress term $-\overline{u_i u_j}$ in the time-averaged momentum equation. This assumption can be removed if we seek a transport equation for the Reynolds stresses directly. An example is the second closure (moment) model. As early as 1951, Rotta (1951) devised a plausible model for the differential equation governing evolution of the tensor that represents turbulent stresses, that is, the Reynolds stress tensor. In essence, the non-local and history effects are somewhat incorporated. Additionally, the model accommodates complicating effects such as streamline curvature, rigid-body rotation, and body forces. For a three-dimensional flow, a second-order closer model introduces seven equations: one for the turbulence scale and six for the components of the Reynolds-stress tensor. Thus, a major drawback is that it involves a large number of equations and complexity.

5.2.3.3 Hybrid methods

The LES and RNG (Re-normalized Group [Yakhot et al., 1992]) theoretical approaches compute the large-scale features (the vortical fluid) from dynamical equations, but time-average the small scales (rapid time, therefore "space filling" and quasi-normal Gaussian), i.e, the dissipation scales. There are challenges to determining how the energy is transferred between the two regimes.

5.3 LARGE EDDY SIMULATION

The large-scale features in turbulent flow are significantly or more directly affected by the flow conditions; recall the turbulent energy cascade discussed in Chapter 4. This, and the fact that they are relatively large in size and fewer in number, suggests that it is relatively less challenging to calculate them. With this knowledge, Smagorinsky (1963), Lilly (1967), Deardorff (1974), and Schumann (1975), among others, pioneered the large-eddy simulation (LES) endeavor.

To divide the large-scale features, which are explicitly resolved via the time-dependent Navier-Stokes equations from the smaller ones, a filter length Δ_{LES} is introduced. This is illustrated in Fig. 5.2, where LES is a

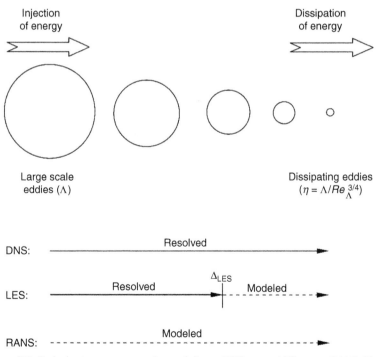

Figure 5.2 Turbulent energy cascade modeling – DNS versus LES versus RANS. *(Created by A.R. Vasel-Be-Hagh).*

balance between DNS where every single eddy is calculated and RANS (such as k-omega and k-epsilon models described in Section 5.2.3) where even the largest eddy is modeled. The word "filter" signifies the removal of scales smaller than the filter length and excludes the direct deduction of them from the Navier-Stokes equations. These smaller and more isotropic eddies, referred to as "subgrid-scale" eddies are modeled. One of the most famous subgrid-scale models is the model proposed by Smagorinsky (1963); Meldi et al. (2011) demonstrated that a filter length of 1/55th the large turbulent scale is the limit for the correct application of the Smagorinsky subgrid scale model.

The whole velocity field, from the largest down to the smallest eddy, described by tensor $U(x, t)$, is required. On the other hand, only the velocity field described by tensor $W(x, t)$, which characterizes scales larger than the filter length, or turbulence-resolved length, is resolved from LES. As such, $W(x, t)$ is not the spatially filtered value of $U(x, t)$. Specifically, $W(x, t)$ is the solution to the LES equations, not as the spatially filtered value of $U(x, t)$, which we denote as $\bar{U}(x, t)$. The question then is whether it is possible to have a perfect LES model such that $W(x, t) = \bar{U}(x, t)$. The answer is no, for $\bar{U}(x, t)$ is a random field whose future evolution is not determined by its current state. In short, the relationship between U and W can only be statistical.

A turbulence model is "complete" if its constituent equations are free from flow-dependent specifications; that is, one flow is distinguished from another solely by the specification of material properties, as well as initial and boundary conditions. The general practice is to generate a computational grid with spacing characterized by the numerical resolution. With this defined, the filter length is specified as proportional to the mesh size. In other words, the turbulence resolution length scale is specified in a flow-dependent, subjective manner. As such, LES is incomplete. LES can be significantly improved, or made "complete," through the use of solution-adaptive meshing. In adaptive LES, the same turbulence-resolution tolerance is met throughout the flow field; for example, 80% of the kinetic energy by adjusting the grid fineness. Piomelli et al. (2015) is one recent example where the filter length is disassociated from the computational grid; it is a dynamic length, which reflects the local, instantaneous turbulence activity.

There remain some fundamental questions about the conceptual foundations of LES, as well as the methodologies and protocols used in its application (Pope, 2004). Nonetheless, there have been serious advances over the years. The fundamental quantity considered in LES is a three-dimensional unsteady velocity field, referred to as the "resolve velocity field," which is intended to represent the larger-scale motions. This resolved velocity field is indeed a three-dimensional, time-dependent random flow field. Nonetheless, as already mentioned, it depends on artificial parameter filter length, mesh spacing, and the numerical method used.

For high-Re, free-shear flows, the transport processes of interest are driven by the resolved, large-scale motions; and there (probably) is a cascade of energy, predominantly from the resolved large scales to the statistically isotropic and universal small scales. As the quantities of interest and the rate-controlling processes are determined by the resolved large scales, LES is likely to work well.

In large-Re, near-wall flows, the shear stress arises from momentum transfer from the outer flow through the boundary layer to the wall. In the viscous near-wall region, the momentum transfer is dominated by the near-wall structures, defining the characteristic length, which scales with the small viscous length scale. Thus, there can be a very large separation between the size of those eddies outside and inside the boundary layer. Consequently, such a flow field may not be properly resolved in high-Re LES, but must instead be modeled (Chapman, 1979). In the pursuit of overcoming this challenge, the detached eddy simulation (DES) emerged (Spallart et al., 1997). In DES, the large eddies are deduced via LES, while the relatively much smaller eddies in the boundary layers or thin shear layers are modeled via RANS (Wilcox, 2006).

5.4 DIRECT NUMERICAL SIMULATION

In direct numerical simulation (DNS), a complete time-dependent, three-dimensional solution of the Navier-Stokes and continuity equations is calculated (Wilcox, 2006). In principle, DNS gives the exact, error-free solution; in practice, however, there are typically some numerical and other forms of errors. As all length scales from the largest to the smallest are explicitly resolved, DNS requires a grid size smaller than the Kolmogorov scale. This, along with the required fine computational time step, makes it a forbidden task for large geometries and/or high Reynolds number problems. It is thus clear that we will continue to depend on good turbulence models to capture the essence of the problem and provide reasonably accurate quantitative values. Whether one is into numerical simulations or

" No matter how hard you try,
you can't solve the problem
unless you choose a correct
model. "

Figure 5.3 Turbulent modeling fever. *(Created by S.P. Mapparapu).*

not, some amount of turbulence modeling is essential for every turbulence investigator; see Fig. 5.3.

Problems

Problem 5.1. A k-u'/λ turbulence model

Evaluate and see if we can create a two-equation turbulence model based on (1) turbulence kinetic energy per unit mass, k, and (2) straining rate, u/λ. Check for the relationships between large length scale l, dissipation rate per unit mass ε, and eddy viscosity ν with k and u/λ.

Problem 5.2. A two-equation turbulence model

Create a two-equation turbulence model that is neither k-ω nor k-ε. Include dimensional analysis and physical arguments to justify your choice of turbulent quantities.

Problem 5.3. A statistical turbulence model

The fluctuations of a simple turbulence can perhaps be described by a Gaussian distribution. Any deviation from this ideal case may be characterized with the help of the third (skewness) and fourth (flatness) moments. Propose such a statistical model.

Problem 5.4. Multiequation turbulence models

Search the literature and evaluate a couple of attempted or existing multiequation models. Propose a new one based on sound reasoning.

REFERENCES

Alammar, K., 2014. Fully-developed turbulent pipe flow with heat transfer using a zero-equation model. Res. J. Appl. Sci. Eng. Technol. 7 (16), 3248–3252.

Baldwin, B.S., Lomax, H., 1978. Thin layer approximation and algebraic model for separated turbulent flows, AIAA Paper 78–257.

Bazargan, M., Mohseni, M., 2012. Algebraic zero-equation versus complex two-equation turbulence modeling in supercritical fluid flows. Comput. Fluids 60, 49–57.

Boussinesq, J., 1877. Essai sur la theorie des eaux courantes. Mem. Pres. Par div. Savants a l'Academie Sci., Paris 23, 1–680.

Boussinesq, J., 1897. Théorie de l'écoulement tourbillonnant et tumultueux des liquides dans les lits rectilignes a grande sectionvol. 2Gautier-Villars, Paris.

Cebeci, T., Smith, A.M.O., 1974. Analyses of Turbulent Boundary Layer, Series in Applied Math & Mechvol. 15Academic Press, New York.

Chapman, D.K., 1979. Computational aerodynamics development and outlook. AIAA J. 17 (12), 1293–1313.

Chen, C.-J., Jaw, S.Y., 1997. Fundamentals of Turbulence Modeling. Taylor & Francis, Bristol.

Chou, P.Y., 1945. On the velocity correlations and the solution of the equations of turbulent fluctuation. Quart. Appl. Math. 3, 38–54.

Davydov, B.I., 1961. On the statistical dynamics of an incompressible fluid. Dokl. Akad. Nauk SSSR 136, 47–49.

Deardorff, J.W., 1974. Three-dimensional numerical study of the height and mean structure of a heated planetary boundary layer. Boundary-Layer Meteorol. 7, 81–106.

Garde, R.J., 2010. Turbulent Flow, third ed. New Age Science, Kent, UK.

Harlow, F.H., Nakayama, P.I., 1968. Transport of turbulence energy decay rate, Los Alamos Sci. Lab., University of California Report LA-3854.

Jeans, J., 1962. An Introduction to the Kinetic Theory of Gases. Cambridge University Press, Cambridge.

Jones, W.P., Launder, B.E., 1972. The prediction of laminarization with a two-equation model of turbulence. Int. J. Heat Mass Transfer V 15, 301–314.

Kolmogorov, A.N., 1942. Equations of Turbulent Motion of an Incompressible Fluid, Izvestia Academy of Sciences, USSR; Physics, vol. 6,1/2,pp. 56–58.

Launder, B.E., Sharma, B.I., 1974. Application of the energy dissipation model of turbulence to the calculation of flow near a spinning disc. Lett. Heat Mass Trans. 1 (2), 131–138.

Launder, B.E., Spalding, D.B., 1972. Lectures in Mathematical Models of Turbulence. Academic Press, London.

Li, C., Li, X., Su, Y., Zhu, Y., 2012. A new zero-equation turbulence model for micro-scale climate simulation. Build. Environ. 47 (1), 243–255.

Lilly, D.K., 1967. The representation of small-scale turbulence in numerical simulation experiments, In: Goldstine, H.H. (Ed.), Proceedings of IBM Scientific Computing Symposium on Environmental Sciences, Yorktown Heights, New York, IBM form No. 320.

Meldi, M., Lucor, D., Sagaut, P., 2011. Is the Smagorinsky coefficient sensitive to uncertainty in the form of the energy spectrum? Phys. Fluids A 23 (12), 1–14.

Ng, K.C., Abdul Aziz, M.A., Ng, E.Y.K., 2011. On the effect of turbulent intensity towards the accuracy of the zero-equation turbulence model for indoor airflow application. Build. Environ. 46 (1), 82–88.

Piomelli, U., Rouhi, A., Geurts, B.J., 2015. A grid-independent length scale for large-eddy simulations. J. Fluid Mech. 766, 499–527.

Pope, S.B., 2000. Turbulent Flows. Cambridge University Press, Cambridge.

Pope, S.B., 2004. Ten questions concerning the large-eddy simulation of turbulent flows. New J. Phys. 6, 35.

Prandtl, L., 1925. Über die ausgebildete turbulenz. ZAMM 5, 136–139.

Prandtl, L., 1945. Über ein neues formelsystem für die ausgebidete turbulenz, Nacr. Akad. Wiss Göttingen, Math-Phys. Kl. 6–19.

Rotta, J.C., 1951. Statistische theorie nichthomogener turbulenz. Zeitschrift fur Physik 129, 547–572.

Russel, J.M., Landahl, M.T., 1984. The evolution of a flat eddy near a wall in an inviscid shear flow. Phys. Fluids A 27, 557–570.

Schumann, U., 1975. Subgrid scale model for finite difference simulations of turbulent flows in plane channels and annuli. J. Comput. Phys. 18 (4), 376–404.

Smagorinsky, J., 1963. General circulation experiments with the primitive equations. 1. The basic experiment. Mon. Weather Rev. 91 (3), 99–164.

Smith, A.M.O., Cebeci, T., 1967. Numerical solution of the turbulent boundary layer equations, Douglas aircraft division report DAC 33735.

Spallart, P.R., Jou, W.-H., Strelets, M., Allmaras, S.R., 1997. Comments on the Feasibility of LES for Wings, and on a Hybrid RANS/LES Approach, Advances in DNS/LES. Greyden Press, Columbus, OH.

Townsend, A.A., 1976. The Structure of Turbulent Shear Flow, second ed. Cambridge University Press, Cambridge.

van Driest, E.R., 1956. On turbulent flow near a wall. J. Aeronaut. Sci. 23 (11), 1007–1011.

Wilcox, D.C., 1993. Turbulence Modeling for CFD. DCW Industries, Inc, La Cañada, California.

Wilcox, D.C., 1998. Turbulence Modeling for CFD, second ed. DCW Industries, Inc, La Cañada, California.

Wilcox, D.C., 2006. Turbulence Modeling for CFD, third ed. DCW Industries, Inc, La Cañada, California.

Yakhot, V., Orszag, S.A., Thangam, S., Gatski, T.B., Speziale, C.G., 1992. Development of turbulence models for shear flows by a double expansion technique. Phys. Fluids A 4 (7), 1510–1520.

CHAPTER 6

Wall Turbulence

It's the little details that are vital. Little things make big things happen.

– John Wooden

Everyone is trying to accomplish something big, not realizing that life is made up of little things.

– Frank A. Clark

Contents

Chapter Objectives

- To recap the boundary-layer concept.
- To discern and differentiate laminar boundary layer from turbulent boundary layer over a flat plate.
- To apply dimensional analysis and deduce the appropriate parameters for different zones within the turbulent boundary layer.
- To differentiate the viscous sublayer from the log region where the law of the wall applies.
- To introduce the defect velocity law region along with the outer layer zone called the wake region.

NOMENCLATURE

BL	Boundary layer
C	Constant, coefficient
C_f	Friction coefficient
CFD	Computational fluid dynamics
f	A particular function
g	A particular function
h	Height
L	Length

Basics of Engineering Turbulence
http://dx.doi.org/10.1016/B978-0-12-803970-0.00006-4

l_{mix} Mixing length
P Time-averaged pressure
Re   Reynolds number
u The fluctuating component of the velocity (in the x direction)
u_{*} Shear or friction velocity, $u_{*} = \sqrt{(\tau_w/\rho)}$
U (Time-averaged) local velocity (in the x direction)
U_{∞} (Time-averaged) free-stream velocity in the x direction
V (Time-averaged) local velocity in the y direction
v The fluctuating component of the velocity in the y direction
W (Time-averaged) local velocity in the z direction
w The fluctuating component of the velocity in the z direction
x Distance in the x (streamwise) coordinate
y Distance in the y (vertical) coordinate
z Distance in the z coordinate

Greek Symbols
δ (Boundary-layer) thickness
δ_d Displacement boundary-layer thickness
δ_m Momentum boundary-layer thickness
δ_v Viscous sublayer thickness
κ Von Kármán constant
μ Dynamic viscosity
v Kinematic viscosity, $v = \mu/\rho$
ξ Normalized distance from the wall, $\xi \equiv y/\delta$
ρ Density
τ Shear
τ_w Wall shear

6.1 INTRODUCTION

In the presence of a solid wall, the flow and thus, the turbulence, is directly influenced. This *wall turbulence* may be divided into two groups. The first involves flows around a rigid body, and the second deals with flows in a space confined by rigid walls. Prior to delving into wall turbulence, a brief overview of boundary layer is due. We limit our discussion to the simplest classical case of wall turbulence, that is, the two-dimensional boundary-layer flow along a flat plate with negligible pressure gradient.

It was Ludwig Prandtl who closed the outstanding gap between *theoretical hydrodynamics*, which evolved from Euler's 1755 (Euler, 1755) equation of motion for a non-viscous (inviscid) fluid, and *hydraulics*, an empirical art developed by practical engineers in 1904 (Prandtl, 1904). At that time, the few idealized inviscid problems solved elegantly in an exact manner were of little use in practice. At the applied end, there was next to no generalization in hydraulics; only cumbersome experiments were conducted for each and every practical undertaking. Prandtl's breakthrough built on the fact that viscosity is only significant within the very thin layer called the boundary or frictional layer, outside of which the flow is essentially inviscid. Thence, it was the very first time the boundary layer was invoked to smoothly wed the inviscid flow to the no-slip condition at the wall via viscosity.

Following White (2011), Fig. 6.1 depicts the development of the boundary layer from a uniform incoming free stream onto a very thin plate. At very low Reynolds numbers (Fig. 6.1a), the viscous effect is especially dominant and hence, the resulting boundary-layer buildup is early and thick, and the gradient of the velocity profile is gradual throughout the entire boundary layer. At high velocities or Reynolds numbers (Fig. 6.1b), the overwhelming inertia pushes the fluid faster and closer to the solid boundary, producing a much thinner boundary layer with a large velocity gradient next to the wall. In other words, the boundary-layer thickness δ is much smaller than L, the streamwise distance from the leading edge. The "gradient" of the velocity is the key parameter in a boundary layer, as it signifies

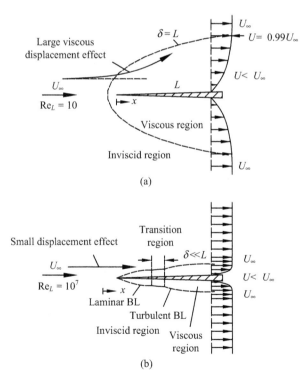

(a)

(b)

Figure 6.1 Boundary layer over a thin plate at (a) very low Re, and (b) very high Re. *(Created by H. Cen).*

"flow shear." Also worth noting is that the value of the pertinent Reynolds number, for which the relevant characteristic length is the distance from the leading edge, starts from zero. It follows that, as can be seen particularly in Fig. 6.1b, the first-formed boundary layer is laminar, and this hurriedly transforms into a turbulent one when the incoming velocity is high. In typical engineering practice, the critical Reynolds number at which the boundary layer undergoes transition from laminar to turbulent is customarily taken as 5×10^5. More importantly, this value largely depends on factors such as pressure gradient, surface roughness, free-stream disturbances, and the workmanship of the leading edge.

6.2 COMMON TYPES OF BOUNDARY-LAYER THICKNESS

The most common *boundary-layer thickness* δ is the distance from the wall where the local velocity U reaches 99% the free-stream value. This boundary-layer thickness, sketched in Fig. 6.2 for flow over a flat surface, is also referred to as the *disturbance thickness* (Pritchard and Mitchell, 2015). The *momentum thickness* δ_m is a measure of the drag imposed on a solid boundary. It can be obtained by performing the momentum integral across the plane where the flow exits. As such, it portrays the notion that the boundary layer retards the fluid so that the momentum flux is less than it would be if the fluid were inviscid; see Fig. 6.2. The third length scale for quantifying the boundary layer is the *displacement thickness* δ_d. As illustrated in Figs 6.2 and 6.3, this displacement thickness is simply the amount of outward (upward in the y direction) shift in the streamlines outside the boundary layer (White, 2011). In other words, for the two-dimensional case considered, the displacement boundary-layer thickness δ_d signifies the height portrayed in Fig. 6.2, which makes the two-hatched areas equal. Mathematically, to satisfy the conservation of mass in the flow direction

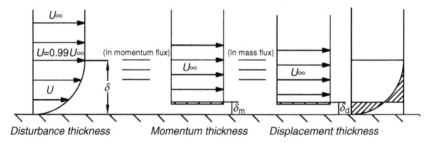

Figure 6.2 Boundary-layer thicknesses – disturbance thickness, momentum thickness, displacement thickness. *(Created by A. Ahmed).*

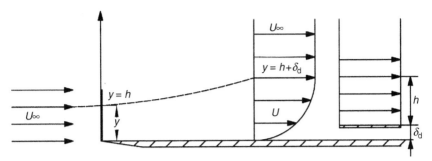

Figure 6.3 Boundary-layer displacement thickness δ_d; the velocity deficit causes the upward shift of height h; the total area below $y = h + \delta_d$ covered by the velocity of the actual and equivalent uniform velocity profile is the same. *(Created by A. Ahmed).*

$$\int_0^h \rho U_\infty z_{unit}\, dy = \int_0^{h+\delta_d} \rho U z_{unit}\, dy \tag{6.1}$$

where h is an arbitrary height above the boundary layer as shown in Fig. 6.3, U_∞ is the free-stream velocity, U is the local velocity, and z_{unit} is the unit width in the z direction.

6.3 FLAT-PLATE BOUNDARY LAYER

Consider the ideal case where an incompressible fluid with uniform velocity flows steadily over a smooth, flat plate as shown in Fig. 6.4. Commencing from the leading edge (unless the Reynolds number is very low, in which case the boundary layer starts ahead of the leading edge as depicted in Fig. 6.1a), a relatively thick (with respect to the streamwise distance from the leading edge) laminar boundary layer grows rapidly. This growth in laminar boundary layer continues until the transition point, where instabilities begin

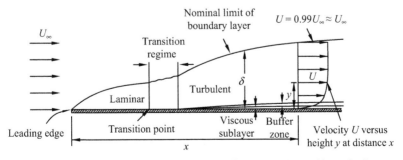

Figure 6.4 The growth of boundary layer along a flat plate. *(Created by H. Cen).*

to be amplified instead of being attenuated by fluid viscosity. These instabilities continue to amplify until the boundary layer becomes fully turbulent,
beyond which the boundary layer grows comparatively slower. While the
whole laminar boundary layer is affected by viscosity, this is not so for the
slower-growing turbulent boundary layer. The turbulent boundary layer
may be *roughly* divided into three sublayers (Wilson, 1989; Schlichting and
Gersten, 2000). Farthest away from the wall is the *outer boundary layer*, where
the velocity profile is relatively uniform; see Fig. 6.4. The flow in this sublayer is characterized by random fluctuating motion and not the fluid viscosity.
Next to the wall is the *viscous sublayer*, also referred to as *viscous wall layer*,
where fluid viscosity plays a dictating role. The *buffer zone* is the matchmaker
that merges these two relatively distinct sublayers together. We shall only
highlight the famed development in the following paragraphs while refer-
ring the readers to standard and specialized fluid mechanics monographs
such as Schetz (1993), Schlichting and Gersten (2000), White (2005, 2011),
and Wilcox (2007) for more detailed coverage.

Let us continue with the steady, two-dimensional, incompressible flow
over a flat plate in the absence of gravity and other forces as depicted in
Fig. 6.4. The continuity equation can be expressed as

$$\frac{\partial U}{\partial x} + \frac{\partial V}{\partial y} = 0 \tag{6.2}$$

where U is the velocity in the streamwise or x direction, and V is the velocity in the direction normal to the solid wall, i.e., the y direction. The corresponding x momentum and y momentum relations are

$$\rho\left(U\frac{\partial U}{\partial x} + V\frac{\partial U}{\partial y}\right) = -\frac{\partial P}{\partial x} + \mu\left(\frac{\partial^2 U}{\partial x^2} + \frac{\partial^2 U}{\partial y^2}\right) \tag{6.3}$$

$$\rho\left(U\frac{\partial V}{\partial x} + V\frac{\partial V}{\partial y}\right) = -\frac{\partial P}{\partial y} + \mu\left(\frac{\partial^2 V}{\partial x^2} + \frac{\partial^2 V}{\partial y^2}\right) \tag{6.4}$$

With the given no-slip solid boundary, inlet, and outlet conditions,
these relations can be solved via today's computational fluid dynamics
(CFD) solvers. In Prandtl's days, however, this was not an option. As mentioned before, there were but a few limited, idealized, and relatively non-
practical situations within which they could be solved exactly. Even with
today's computational power, it is often uneconomical and impractical to

completely resort to heavy-duty CFD for many engineering problems. In other words, simplified approaches which depict the underlying physics are timelessly important.

Following Prandtl, at high Re, the shear layer is very thin; hence, the following approximations can be applied (Wilson, 1989; White, 2011).

1. The cross-stream velocity is much smaller than the streamwise counterpart

$$V \ll U \tag{6.5}$$

2. The rate of change in velocity in the streamwise direction is significantly less than that in the cross-stream direction

$$\frac{\partial U}{\partial x} \ll \frac{\partial U}{\partial y} \tag{6.6}$$

$$\frac{\partial V}{\partial x} \ll \frac{\partial V}{\partial y} \tag{6.7}$$

3. Since Re $\gg$ 1, we have

$$\mathrm{Re}_x = \frac{Ux}{\nu} \gg 1 \tag{6.8}$$

Applying these approximations to Eq. (6.4), we see from the order of magnitude perspective that

$$\text{small} + \text{small} = -\partial P / \partial y + \text{very small} + \text{small} \tag{6.9}$$

or

$$\partial P / \partial y \approx 0 \tag{6.10}$$

that is, $P \approx P(x)$ only. Furthermore, applying Bernoulli's equation to the outer inviscid flow, we get

$$\partial P / \partial x = dP / dx = -\rho U_\infty \, dU_\infty / dx \tag{6.11}$$

where U_∞ is the free-stream velocity, as compared to the local velocity U.

It is thus clear that the three equations of motion may be simplified into Prandtl's two boundary layer equations. Specifically, for two-dimensional, incompressible, steady flow, the flow continuity can described by Eq. (6.2). The corresponding momentum along the wall

$$U\frac{\partial U}{\partial x}+V\frac{\partial V}{\partial y}\approx U_{\infty}\frac{dU_{\infty}}{dx}+\frac{1}{\rho}\frac{\partial \tau}{\partial y} \tag{6.12}$$

where, from Chapter 2, for laminar flow

$$\tau = \mu\frac{\partial U}{\partial y} \tag{6.13}$$

and for turbulent flow

$$\tau = \mu\frac{\partial U}{\partial y} - \overline{\rho uv} \tag{6.14}$$

The no-slip boundary condition at the wall implies that at $y = 0$, both U and V are zero. At the boundary layer and beyond, $U = U_{\infty}$. With these boundary conditions, the above four equations, Eqs (6.2) and (6.12–6.14) can be used to solve for the velocity field, $U(x, y)$ and $V(x, y)$.

It is noteworthy that boundary layers are self-similar in the sense that they can be collapsed in a general, non-dimensional manner. We shall proceed along this line of thought as we look at the various formulations of universal similarities between velocity profiles.

6.3.1 Laminar Boundary Layer

The general growth of the laminar boundary layer with respect to the streamwise distance from the leading edge for the constant U_{∞} ($dU_{\infty}/dx = 0$) case was first elegantly solved by Prandtl's student Blasius (1908). The approximation

$$\frac{\delta}{x} \approx \frac{5.0}{\mathrm{Re}_x^{1/2}} \tag{6.15}$$

is valid for $10^3 < \mathrm{Re}_x < 10^6$. The corresponding *skin friction coefficient*

$$C_f = \frac{2\tau_w}{\rho U_{\infty}^2} = \frac{0.664}{\mathrm{Re}_x^{1/2}} \tag{6.16}$$

where τ_w is the wall shear.

6.3.2 Transition to Turbulent

Depending on factors such as surface roughness, workmanship of the leading edge, and free-stream turbulence level, transition into a turbulent boundary layer over a smooth surface occurs at Re_x of approximately 5×10^5. The increase in boundary thickness is hastened when the flow becomes turbulent;

see Fig. 6.4. Schubauer and Skramstad (1943) studied the effect of free-stream turbulence level on the laminar-to-turbulent transition over a flat plate in a wind tunnel. They found that the effect of free-stream turbulence diminishes when its intensity falls below 0.1%, and it has to be greater than 0.2% to cause a notable influence on the critical Reynolds number. The value of the critical Reynolds number Re_c decreases from 3×10^6 to roughly 10^6 when the relative turbulence intensity is increased to 0.5%, and, according to Dryden (1947), it further reduces to around 10^5 at a relative turbulence intensity of 3%.

6.3.3 Turbulent Boundary Layer

We have been acquainted with the fact that the turbulent boundary can *roughly* be divided into three sublayers at the onset of Section 6.3 with the help of Fig. 6.4. These three sublayers have different velocity distributions and hence, require three equations to describe them, as opposed to only one needed for the laminar boundary layer. Let us follow the gentle approach taken by Elger et al. (2013) as we delve into these sublayers and refine the divisions or categorizations. Before we proceed with that, it is worth presenting a similar but somewhat different approximate division of the turbulent boundary layer as illustrated in Fig. 6.5 (Wilson, 1989). There is a general consensus regarding the next-to-the-wall, viscosity-dominating, viscous sublayer. The layer farther out is referred to as the *inner inertial layer*, which is a transition or buffer region separating the viscous sublayer from the outer inertial layer. This inner inertial layer overlaps slightly with the outer inertial layer. In this buffer region, the effects of viscosity and turbulence inertia are of the same order of magnitude in the inner inertial layer. As such, viscosity comes into play indirectly via the shear velocity (to be defined in the next paragraph). This shear velocity signifies the intensity of turbulence and it scales with the local velocity U, whereas the appropriate length scale is the viscous length ν/u_* (also explained in the next paragraph). The outermost layer where the free-stream velocity starts to feel the

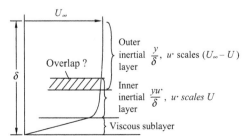

Figure 6.5 Rough divisions of plane-turbulent boundary layer. *(Created by H. Cen).*

effect of the wall and slows down is called the *outer inertial layer*. It is obvious that the boundary-layer thickness is the appropriate choice for normalizing the distance from the wall, and that the shear velocity is related to the free-stream velocity, as the local velocity is very close to it. Moreover, the overall dynamics in this inertial layer are independent of fluid viscosity, just as the large-scale spectral dynamic of turbulence is. In other words, an inertial sublayer in wall-bounded shear flows bears resemblance to the inertial sub-range in the turbulence energy spectrum.

Recall that fluid viscosity dominates in the small region right next to the wall, and hence, the flow is essentially laminar. Accordingly, the *viscous sublayer* has also been called the *laminar sublayer*. We note, however, that the latter designation may not be most appropriate, as the flow in the viscous sublayer is not strictly laminar. Nonetheless, the thin viscous sublayer does behave like Couette flow; where the laminar viscous fluid flow is between two parallel plates, one of which is moving. For Couette flow, the velocity gradient is a constant, and therefore, the shear. By analogy, in the viscous sublayer, the shear stress τ is basically constant and is equal to τ_w, the shear stress at the wall. Thus

$$dU/dy = \tau_w/\mu \tag{6.17}$$

where μ is the dynamic viscosity of the fluid. Upon integration, we have

$$U = \tau_w y/\mu \tag{6.18}$$

Multiplying the right hand side by ρ/ρ gives

$$U = \frac{\tau_w/\rho}{\mu/\rho} y \tag{6.19}$$

This can be rewritten as

$$\frac{U}{\sqrt{\tau_w/\rho}} = \frac{\sqrt{\tau_w/\rho}}{\nu} y \tag{6.20}$$

We note that $\sqrt{(\tau_w/\rho)}$ has the dimension of velocity and therefore is dubbed *shear velocity* or *friction velocity*, $u_\star$. With this shear velocity, we can express Eq. (6.20) in the well-known nondimensional form

$$\frac{U}{u_\star} = \frac{y}{\nu/u_\star} \tag{6.21}$$

which is the standard form used to describe the velocity distribution in the viscous sublayer. The denominator on the right hand side v/u_* is the *viscous length*; that is, it is a characteristic length dictated by the fluid viscosity. Also worth highlighting is that the shear velocity u_* signifies a characteristic level of turbulence in the boundary layer.

At this point, we take a break from dimensional analysis and more or less emulate Cowen (2015) and Elger et al. (2013) by looking at the problem at hand. We have an incompressible uniform flow over a smooth, flat plate at some high Reynolds number. The well-developed boundary layer grows very slowly in the streamwise direction and thus, as per Eqs (6.5) and (6.6), $\partial/\partial x \ll \partial/\partial y$. The boundary conditions are the no-slip condition, $U(y = 0) = 0$, and the wall shear as per discussion on viscous sublayer; the wall shear $\tau_{xy}(y = 0) = \tau_{yx}(y = 0) = \tau_w$. From Eq. (6.14), we have

$$\tau_w = \mu \frac{\partial U}{\partial y} - \overline{\rho u v} \tag{6.22}$$

Therefore, this wall shear is what we wish to solve, keeping in mind that the yet-to-be-utilized boundary condition is $U(y = 0) = 0$. Let us examine the broader boundary-layer region where viscosity is not as overwhelming as it is in the viscous sublayer in overshadowing fluid inertia, keeping all instabilities or turbulent fluctuations in check. A faster-moving fluid element farther out in the boundary layer (when tossed, due to random turbulent fluctuations, into the lower velocity region closer to the wall) tends to accelerate the sluggish fluid, and vice versa. This turbulence–induced and significantly enhanced momentum transport phenomenon, as discussed in Chapters 2 and 5, can be regarded as the outcome of applying effective Reynolds stresses to the otherwise languorous fluid. The faster-moving fluid particle may be viewed as a flying stone as shown in Fig. 6.6, which jerks the unhurried wagon forward as it plunges inside.

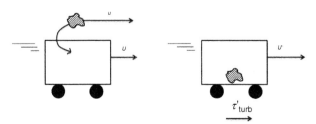

Figure 6.6 Turbulent momentum transport enhanced by Reynolds stresses. *(Created by A. Goyal).*

Here is what we have gathered so far concerning the broader boundary-layer region, which is outside of the viscous sublayer. As the surface is smooth, the surface roughness height, h_r, is not a factor and hence, $U = U(y, \rho, \nu, \tau_w, \delta, U_\infty)$. Therefore, nondimensionally we have

$$u^+ = U(y)/u_\star \tag{6.23}$$

where the friction velocity $u_\star = \sqrt{(\tau_w/\rho)}$. The (normal) distance from the wall y can be normalized using the viscous length $\nu/u_\star$ to give the famous y^+, that is

$$y^+ = u_\star y/\nu \tag{6.24}$$

In addition, we can normalize the larger distance from the wall by the boundary layer thickness

$$\xi = y/\delta \tag{6.25}$$

The broader boundary-layer region is outside of the viscous sublayer so that the flow structure is affected by ν only through its influence on $u_\star$. Yet, we are close enough to the wall such that y^+, not the boundary layer thickness δ or ξ, is the relevant length variable. This assumption allows us to avoid the difficulty of having to simultaneously deal with two length scales, δ and $\nu/u_\star$.

To move further with the analysis of the broader boundary layer region, we can conjure the eddy viscosity model covered in Chapter 5 and Eq. (6.22), and obtain

$$-\overline{\rho uv} = \nu_{turb} \frac{\partial U}{\partial y} \tag{6.26}$$

Realizing that the eddy size probably varies with distance from the wall in the boundary layer, Prandtl (1925) conjectured that l_{mix} adjusts itself in proportion to the distance y; subsequent development led to

$$l_{mix} = \kappa y \tag{6.27}$$

where the von Kármán constant, $\kappa = 0.41$. This, along with the original mixing length concept

$$u \approx l_{mix} \, dU/dy \tag{6.28}$$

we have

$$\tau_{\mathrm{w}} = \rho \kappa^2 \, y^2 \left(\frac{dU}{dy} \right)^2 \tag{6.29}$$

After taking the square root, this can be rearranged into

$$dU = [\sqrt{(\tau_{\mathrm{w}}/\rho)}/\kappa] \, dy/y \tag{6.30}$$

Substituting for $u_\star$, we have

$$dU/u_\star = (1/\kappa) \, dy/y \tag{6.31}$$

Integrating we get

$$U/u_\star = (1/\kappa) \ln y + C_{\mathrm{ln}} \tag{6.32}$$

where C_{ln} is the log region constant. Accordingly, there is a range of boundary layer, which we may be able to describe using this logarithmic expression. This region is called the *log region*, with a generally accepted C_{ln} value of 5.5 for a smooth wall. Henceforth we can redefine the layers within the boundary layer as depicted in Fig. 6.7 (based on Elger et al., 2013).

The term, *law of the wall*, has been adopted to describe the logarithmic velocity distribution region, which was first discovered by von Kármán (1930). Following Elger et al. (2013), the logarithmic line has been extended into the viscous sublayer as shown in Fig. 6.8. It is obvious that the

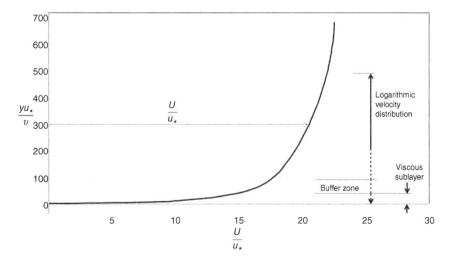

Figure 6.7 Velocity profile in flat-plate turbulent boundary layer. *(Created by A. Goyal).*

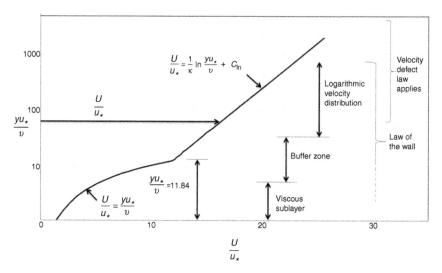

Figure 6.8 Logarithmic velocity profile in a turbulent boundary layer over a flat plate. *(Created by A. Goyal).*

viscous sublayer does not follow the law of the wall, which was derived under the assumption that viscosity does not directly come into play except via the friction velocity. Just as importantly, we notice in Fig. 6.8, and also in Fig. 6.7, that the logarithmic region does not extend outward to the outer edge of the boundary layer. This outer edge region is called the *velocity defect law* because there is some unmistakable velocity deficit with respect to the free-stream velocity.

Figure 6.9 is a schematic detailing various velocity distribution regions (modified after Cowen, 2015). We note that as the velocity defect zone is near the outer edge of the boundary layer, the appropriate length is the boundary-layer thickness δ or nondimensionally, ξ. In this velocity defect law zone the (time-averaged) local velocity U approaches the free-stream value U_∞ as y approaches δ. Also noteworthy is that the law of the wall, expressed by Eq. (6.32), has a logarithmic portion for $y^+ \geq 30$. The extent of the logarithmic region, or the "logarithmic sublayer," depends on the overall flow parameters such as $\mathrm{Re}_\delta \equiv U\delta/v$ and the pressure gradient. Typically, the log region occupies about one-tenth of the boundary layer thickness.

Let us analyze the area spanned by the logarithmic velocity distribution region to the outer extent of the velocity defect region by invoking the following assumptions:

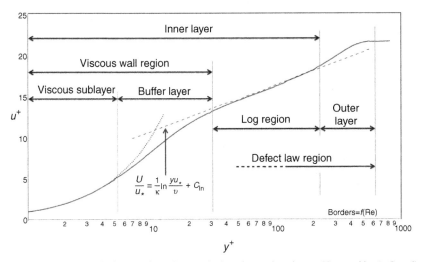

Figure 6.9 The multifarious flat-plate turbulent boundary layer. *(Created by A. Goyal).*

1. The velocity defect law applies in a zone close enough to the wall that the velocity defect ΔU is directly proportional to the friction velocity u_*; that is, not U_∞.
2. There is at least a small zone of overlap where both law of the wall and velocity defect law apply.

Our form is

$$(U_\infty - U)/u_* \equiv g(y/\delta) \tag{6.33}$$

Take $\left.\dfrac{\partial}{\partial y}\right|_x$ using $\xi \equiv y/\delta$ so that $\left.\dfrac{\partial g}{\partial y}\right|_x = \left.\dfrac{\partial g}{\partial \xi}\right|_x \left.\dfrac{\partial \xi}{\partial y}\right|_x$, we have

$$-\frac{1}{u_*}\left.\frac{\partial U}{\partial y}\right|_x = \frac{1}{\delta}\frac{dg}{d\xi} \tag{6.34}$$

From Eq. (6.31) we can express the law of the wall as

$$\frac{\partial U}{\partial y} = \frac{u_*}{\kappa y} \tag{6.35}$$

In the *overlap zone*, both *law of the wall* and *velocity defect law* apply; hence, we substitute Eq. (6.35) into Eq. (6.34) to get

$$-\frac{1}{u_\star}\left(\frac{u_\star}{\kappa y}\right)=\frac{1}{\delta}\frac{dg}{d\xi} \tag{6.36}$$

Rearranging and recalling that $\xi = y/\delta$, we get

$$\frac{dg}{d\xi}=-\frac{1}{\kappa\xi} \tag{6.37}$$

This can be integrated to give

$$g=-\frac{1}{\kappa}\ln\xi+C_{VD} \tag{6.38}$$

where C_{VD} is the velocity defect constant. Therefore

$$\frac{U_\infty-U}{u_\star}=-\frac{1}{\kappa}\ln\left(\frac{y}{\delta}\right)+C_{VD} \tag{6.39}$$

We note that von Kármán's constant appears in both law of the wall and velocity defect law. The velocity defect law holds over most of the boundary layer except the viscous sublayer, which generally is a very thin region, $y^+\approx30$, corresponding typically to about 1% of δ at high Re.

By noting that both the law of the wall and the velocity defect law apply in the overlapping region, we can relate the two coefficients C_{ln} and C_{VD}. In the region of overlap we can write

$$\frac{U_\infty}{u_\star}-\left[\frac{1}{\kappa}\ln\left(\frac{yu_\star}{v}\right)+C_{ln}\right]=-\frac{1}{\kappa}\ln\left(\frac{y}{\delta}\right)+C_{VD} \tag{6.40}$$

But

$$\ln\left(\frac{yu_\star}{v}\right)=\ln\left(\frac{y}{\delta}\frac{\delta u_\star}{v}\right)=\ln\left(\frac{y}{\delta}\right)+\ln\left(\frac{\delta u_\star}{v}\right) \tag{6.41}$$

Therefore

$$\frac{U_\infty}{u_\star}-\frac{1}{\kappa}\ln\left(\frac{y}{\delta}\right)-\frac{1}{\kappa}\ln\left(\frac{\delta u_\star}{v}\right)-C_{ln}=-\frac{1}{\kappa}\ln\left(\frac{y}{\delta}\right)+C_{VD} \tag{6.42}$$

Defining

$$C_f \equiv \frac{\tau_w}{\rho U^2 \delta/2}, \quad \sqrt{\frac{C_f}{2}} \equiv \frac{u_\star}{U_\delta}, \quad \mathrm{Re}_\delta \equiv \frac{U_\infty \delta}{\nu} \tag{6.43}$$

we get

$$C_{VD} = \sqrt{\frac{2}{C_f}} - \frac{1}{\kappa} \ln\left(\sqrt{\frac{C_f}{2}} \mathrm{Re}_\delta \right) - C_{ln} \tag{6.44}$$

Beyond the outer extent of the law of the wall, the flow behaves like a wake; hence, it is called the *wake region* (Coles, 1956) or the *outer layer*. This is portrayed in Figs 6.8 and 6.9, which show that the logarithmic velocity distribution region does not meet with the boundary layer; that is, it is buffered from the free stream by the outer layer or wake region. Like wake flows, shear stress acts to produce a velocity deficit in this outer region. For the negligible pressure gradient flat plate case considered

$$\frac{U_\infty - U}{u_\star} = 9.6\left(1 - \frac{y}{\delta} \right)^2 \tag{6.45}$$

for $y/\delta > 0.15$ (Hama, 1954). According to White (1974), Clauser (1956) suggested that the correct scaling parameter is

$$\Delta \equiv \int_0^\infty \frac{U_\infty - U}{u_\star} dy = \delta_d \sqrt{\frac{2}{C_f}} \tag{6.46}$$

where the displacement thickness.

$$\delta_d \equiv \int \left(1 - \frac{\overline{U}}{\overline{U_\delta}} \right) dy$$

Problems
Problem 6.1. Boundary-layer thicknesses
Clearly and concisely deduce the momentum, displacement, and boundary-layer thicknesses for a uniform flow over a smooth, flat plate at $\mathrm{Re}_x = 3 \times 10^2, 3 \times 10^5$, and 3×10^6. Include good illustrations.

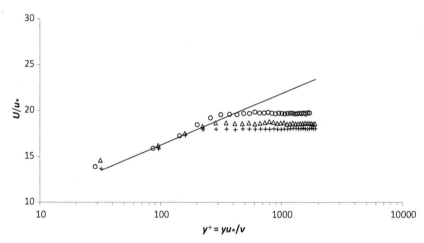

Figure 6.10 Logarithmic profiles of the flat-plate boundary layer. Crosses, triangles, circles signify, respectively, 20, 30, 40 cm from the leading edge. *(Created by F. Fouladi).*

Problem 6.2. Boundary-layer profile

Figure 6.10 shows the logarithmic velocity profiles of the turbulent boundary layer over a flat plate at different streamwise locations, $x = 20, 30$, and 40 cm; where x is the streamwise distance from the leading edge of the plate (Fouladi et al., 2015).

1. Which profile corresponds to which streamwise distance? Explain.
2. How do you expect an enhanced free-stream turbulence to affect these profiles? Explain.

Problem 6.3. Energy dissipation for transition to turbulence

One criterion for determining the point of laminar-turbulent transition is that transition occurs when turbulent flow has a higher rate of entropy production than laminar flow. This will occur when the dissipation rate in the turbulent flow exceeds that of laminar flow.

1. For fully developed flow in a smooth, round pipe, derive expressions for the dimension-less dissipation rate $\varepsilon D^4 / \nu^3$ for laminar and turbulent flow. The dissipation ε is the average over the flow cross-section and can be estimated by a mechanical energy balance using the entire pipe of length L as a control volume. For turbulent flow, you can use the Blasius (1910) resistance law friction factor

$$f = 0.316 / \mathrm{Re}^{1/4} \tag{6.47}$$

for smooth pipe where $\mathrm{Re} = UD/\nu$.
What is the value of Re at laminar-turbulent transition?

2. The Blasius equation is for fully developed turbulent flow. In the laminar-turbulent transition region, a better equation may be the Churchill's (1977) equation, which for smooth pipes is

$$f = 8\left\{ \left(\frac{8}{\mathrm{Re}}\right)^{12} + \left[\left(2.21\times\ln\left(\frac{\mathrm{Re}}{7}\right)\right)^{16} + \left(\frac{37530}{\mathrm{Re}}\right)^{16} \right]^{-3/2} \right\}^{1/2} \quad (6.48)$$

At high Re, this smooth pipe equation reduces to

$$(8\,f)^{-0.5} = 2.21\ln(\mathrm{Re}/7) \quad (6.49)$$

Show that for fixed transition Reynolds number, the dissipation equality simply reduces to $f_{\text{laminar}} = f_{\text{turbulent}}$ at $\mathrm{Re}_{\text{transition}}$. Find the laminar-transitional and transitional fully turbulent flow Reynolds numbers using Eq. (6.48). A graphical plot is helpful.

Problem 6.4. Values of coefficients $C_{\ln}$ and C_{VD}

Deduce the values of coefficients $C_{\ln}$ and C_{VD} for pipe flows, and compare these values with turbulent flow over a flat plate.

REFERENCES

Blasius, H., 1908. Grenzschichten in Flüssigkeiten mit kleiner Reibung. Z. Math. Physik Bd. 56, 1–37, English translation in NACA-TM-1256.

Blasius, H., 1910. Laminare Strömung in Kanälen wechselnder Breit. Z. Math. Physik Bd. 58, 225–233.

Churchill, S.W., 1977. Friction factor equation spans all fluid flow regimes. Chem. Eng. 7, 91–92.

Clauser, F.H., 1956. The turbulent boundary layer. Adv. Appl. Mech. 4, 1–51.

Coles, D., 1956. The law of the wake in the turbulent boundary layer. J. Fluid Mech. 1, 191–226.

Cowen, C.A., 2015. Experimental Methods in Fluid Dynamics (CEE 4370/6370 / MAE 6270), Available from: http://ceeserver.cee.cornell.edu/eac20/cee637/handouts/TURBFLOW_L9.pdf (accessed 11.06.2015.).

Dryden, H.L., 1947. Some recent contributions to the study of transition and turbulent boundary layers, NACA Technical Notes 1168.

Elger, D.F., Williams, B.C., Crowe, C.T., Roberson, J.A., 2013. Engineering Fluid Mechanics, tenth ed. Wiley, USA.

Euler, L., 1755. Principes généraux du mouvement des fluides, MASB, 11: 274–315 (printed in 1757. Also in Opera Omnia, Ser. 2, 12, 54–91.

Fouladi, F., Henshaw, P., Ting, D.S-K., 2015. Effect of a triangular rib on a flat plate boundary layer. J. Fluids Eng. 138, 011101.

Hama, F.R., 1954. Boundary-layer characteristics for smooth and rough surfaces. Trans. Soc. Naval Arch. Marine Engrs. 62, 333–358.

Prandtl, L., 1904. Über flüssigkeitsbewegung bei sehr kleiner reibung. Verhandlungen des dritten internationalen Mathematiker-Kongresses, Heidelberg, Germany, pp. 484–491.

Prandtl, L., 1925. Über die ausgebildete turbulenz. ZAMM 5, 136–139.

Pritchard, P.J., Mitchell, J.W., 2015. Fox and McDonald's Introduction to Fluid Mechanics, ninth ed. Wiley, USA.

Schetz, J.A., 1993. Boundary Layer Analysis. Prentice Hall, Englewood Cliffs.

Schlichting, H., Gersten, K., 2000. Boundary Layer Theory, eighth ed. Springer, Berlin.

Schubauer, G.B., Skramstad, H.K., 1943. Laminar boundary-layer oscillations and stability of laminar flow, National Bureau of Standards, Research Paper 1772 (Also in J. Aero. Sci., 14(2), 69–78, 1947).

von Kármán, Th. 1930, Mechanische Ähnlichkeit und Turbulenz, Nachrichten von der Gesellschaft der Wissenschaften zu Göttingen, Fachgruppe 1 (Mathematik) 5: 58–76 (also as: Mechanical Similitude and Turbulence, Tech. Mem. NACA, no. 611, 1931.

White, F.M., 1974. Viscous Fluid Flow. McGraw Hill, New York.

White, F.M., 2005. Viscous Fluid Flow, third ed. McGraw Hill, New York.

White, F.M., 2011. Fluid Mechanics, seventh ed. McGraw Hill, New York.

Wilcox, D.C., 2007. Basic Fluid Mechanics, third ed. DCW Industries, San Diego.

Wilson, D.J., 1989. Mec E 632: Turbulent Fluid Dynamics, Lecture Notes. University of Alberta, Edmonton.

CHAPTER 7

Grid Turbulence

I look upon experimental truths as matters of great concernment to mankind.

– Robert Boyle

Contents

Chapter Objectives

- To explain homogeneity and isotropy in the context of turbulent flow.
- To investigate flow turbulence downstream of a grid.
- To recognize the different grid-generated turbulence regions.
- To assess the power-law decay region in detail.
- To quantify the key turbulence parameters in terms of distance downstream of the grid.

NOMENCLATURE

a	Exponent
A	Constant coefficient
b	Exponent
B	Proportionality constant
C	(Turbulence decay) coefficient
D	Diameter (of the hole of the perforated plate)
E	Spectral density
F	Flatness factor
k	Turbulence kinetic energy per unit mass, or wave-number

Basics of Engineering Turbulence
http://dx.doi.org/10.1016/B978-0-12-803970-0.00007-6

k_1	Streamwise wave-number
l	Large length scale
M	Mesh or grid size
n	Exponent, frequency
OPP	Orificed, perforated plate
p	The fluctuating component of pressure
r	Space, spatial dimension
P	Time-averaged pressure
q	Two times the square root of kinetic energy per unit mass
Re	Reynolds number
S	Skewness factor
SHPP	Straight-hole perforated plate
t	Time
Tu	Percentage turbulence intensity
u	The fluctuating component of the velocity (in the x-direction)
u_η	Kolmogorov velocity scale
U	Velocity (in the x-direction)
v	The fluctuating component of velocity in the y-direction
V	Time-averaged velocity in the y-direction
w	The fluctuating component of velocity in the z-direction
W	Time-averaged velocity in the z-direction
x, y, z	Cartesian coordinates (x is the streamwise direction)
η	Kolmogorov length
Λ	Integral length
λ	Taylor microscale
ν	Viscosity
ε	Dissipation rate
$\forall$	Volume

7.1 INTRODUCTION

Turbulence generated by a grid has a special place in the heart of turbulent flow. Without elaborating too much, the idea of empirical flow turbulence was initiated via rigorous, systematic grid turbulence experimentations at a time when the concept of turbulent flow was still being formulated. These groundbreaking pursuits include Simmons and Salter (1934), Taylor (1935), Synge and Lin (1943), Lin (1948), and von Kármán and Lin (1949). The extended series of grid turbulence studies conducted by turbulence giants and forefathers Batchelor and Townsend (1947, 1948a, 1948b) are particularly worth mentioning here. In this chapter, we first briefly review the ideal (i.e., impossible) states of isotropic and homogeneous turbulence. As presumably the cleanest and simplest flow turbulence that we can generate, grid turbulence typifies isotropic turbulence. A recent study by Djenidi et al. (2013) showed that the temporal and spatial averages merge at about 20 hole diameters downstream of the grid. With the temporally stationary and spatially homogeneous turbulence in the cross-stream directions, this equality supports the ergodic hypothesis in grid turbulence. As such, this verifies that grid turbulence is indeed homogeneous and isotropic. The rest of this chapter is devoted to grid turbulence.

7.2 HOMOGENEOUS AND ISOTROPIC TURBULENCE

The simplest form of turbulence is the quintessential isotropic turbulence, in which all of its properties are invariant with respect to direction; that is, it is exactly the same in every direction. Among other criteria, the joint probability distribution of the velocities at any arbitrarily chosen n points in space is invariant under arbitrary rotations of the configuration in isotropic turbulence (Batchelor, 1953)

$$\left|\overline{uv}\right| = \left|\overline{vw}\right| = \left|\overline{uw}\right| \tag{7.1}$$

Accordingly, a minimum number of quantities and correlations are needed to describe the structure and behavior of isotropic turbulence (Hinze, 1975).

Homogeneity in flow turbulence typically implies that the associated flow properties do not vary spatially; specifically, the turbulence is the same everywhere in the flow field. We can note that isotropy exists only when the turbulence is already homogeneous, as a nonhomogenous turbulence would show a preference for certain directions. In other words, a change in turbulence with spatial variation requires a certain degree of anisotropy. More interestingly, Hinze (1975) pointed out that a spatially homogeneous turbulence cannot be stationary; therefore, a homogeneous turbulent flow field must be a decaying turbulent field. This is quite hunky-dory as far as grid turbulence is concerned, for the turbulence downstream of the development region is indeed decaying. As such, homogeneity is only true in the cross-stream direction. However, because this decay usually happens slowly, the assumption of the homogeneity of the turbulence is valid for most purposes.

Another outcome or condition of isotropic turbulence is that the probability density function of the fluctuating velocity follows the Gaussian distribution. As discussed in Chapter 3, the corresponding skewness and flatness factors are zero and three, respectively; that is

$$S = \frac{\overline{u^3}}{\overline{u^2}^{3/2}} = 0 \tag{7.2}$$

$$F = \frac{\overline{u^4}}{\overline{u^2}^2} = 3$$

$$\text{(7.3)}$$

Tresso and Munoz (2000) commented that the covariances (cross correlations of u, v, and w) may be the preferred method for experimentally identifying the homogeneous, isotropic flow region, as the skewness is remarkably sensitive to small flow perturbations.

It must be emphasized that there is no real life situation in which turbulence is perfectly isotropic. Local isotropy, on the other hand, can often be assumed to enable the characterization and understanding of turbulence. In the case of fully developed flow turbulence with a well-defined cascade of eddies, the higher-frequency, smaller eddies are quite isotropic, as elucidated in Chapter 4.

7.3 CHARACTERISTICS OF GRID TURBULENCE

Turbulence may be defined as a spatially complex distribution of vorticity, which advects itself in a chaotic manner. The velocity field is determined at any given moment by the fluid's vorticity distribution, in accordance with the Biot-Savart law. Thus, creation of turbulence requires the generation of vorticity; a grid is commonly and successfully used to accomplish this purpose.

We may divide the flow downstream of a grid into four regions:

1. The initial "developing" region
2. The "simple" or "power-law" decay region
3. The "dominating large-scale" region
4. The "final period of decay" region

The eddying motion, along with the corresponding fluctuating velocity, is sketched in Fig. 7.1.

Taking the lead from the innovative work that preceded them, Liu et al devised a unique orificed, perforated plate (OPP), as shown in Fig. 7.2, to generate extremely clean wind tunnel turbulence (Liu et al., 2004, 2007; Liu and Ting, 2007). The acquired data are decomposed into mean and fluctuating velocities, as plotted in Fig. 7.3. The measurements were taken right along the center line of the plate, which is also the center line of the middle hole. We see that the mean or time-averaged velocity maxes out immediately after the OPP, while the fluctuating velocities peak between three and four OPP hole diameters downstream. Accordingly, the initial turbulence-developing or generating region for this OPP ends within approximately five hole diameters downstream. This is significantly shorter than conventional and finite thickness grids.

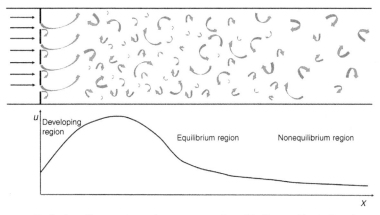

Figure 7.1 Turbulent flow regimes downstream of a grid. *(Created by A. Goyal).*

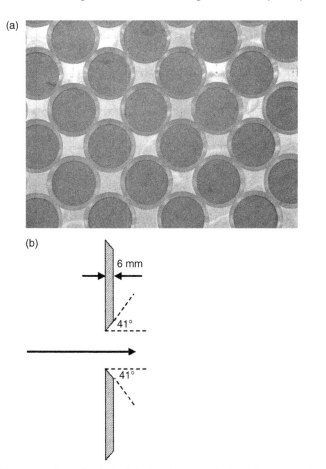

Figure 7.2 The orificed, perforated plate (a) close-up photo, (b) cross-sectional view. *(Taken/created by R. Liu).*

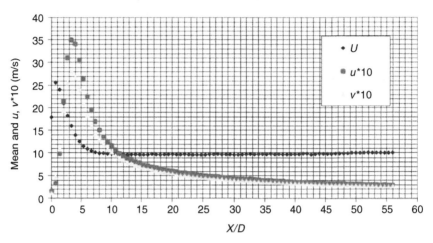

Figure 7.3 Mean and fluctuating velocities downstream of an orificed, perforated plate. *(Created by R. Liu).*

To be thorough, a sample time-averaged local velocity deviation from the center line value of the OPP's turbulent flow is presented in Fig. 7.4. We see that the largest deviation of the local velocity from the center line value is less than 4%; that is, all values are within 97% and 104% of the center value. To that end, the time-averaged velocity over the considered cross section is clearly homogeneous.

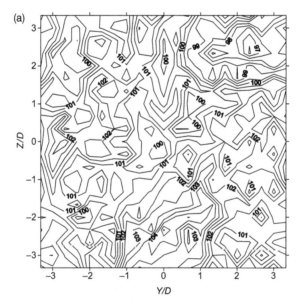

Figure 7.4 Percentage of local time-averaged velocity of the center line value of 10.8 m/s at (a) 20D, (b) 60D, (c) 100D downstream of the OPP. *(Created by R. Liu).*

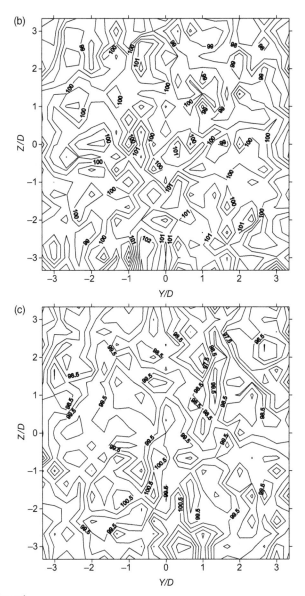

Figure 7.4 (*cont.*)

The corresponding streamwise turbulence fluctuating velocity of the OPP turbulent flow is depicted in Fig. 7.5. The OPP-generated turbulence limns a uniform profile at all three studied cross sections with a maximum departure from the average of no more than 5%, even at just $20D$ downstream. This is remarkable considering the fact that conventional grid-generated turbulent flow typically does not become homogeneous

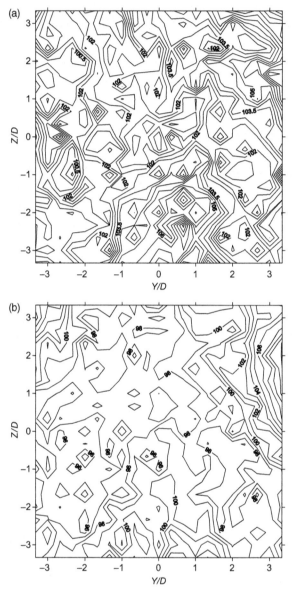

Figure 7.5 Percentage of streamwise root-mean-square fluctuating velocity of that at the center line value at (a) 20*D* with a center line value of 0.64 m/s, (b) 60*D* with a center line value of 0.33 m/s, (c) 100*D* with a center line value of 0.26 m/s, downstream of the OPP. *(Created by R. Liu).*

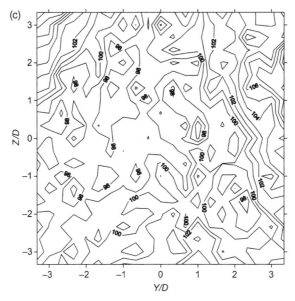

Figure 7.5 (*cont.*)

until 30–40 diameters downstream (Stewart and Townsend, 1951; Portfors and Keffer, 1969). This implies that we may better and more easily approach isotropic turbulence using an OPP, presumably because it promotes three-dimensional flow immediately behind the orifice (Mi et al., 2001). Also worth noting is the rapid decay of the turbulence, as indicated by the fluctuating velocity that rapidly decreases over distance.

Figure 7.6 shows the corresponding covariance or correlation between u and v of the OPP turbulence. We see that the $\overline{uv}$ is literally zero and shows clean and isotropic turbulence. In other words, according to Fig. 3.11 and the associated discussion in Chapter 3, u and v behave as independent variables. Recall that independent variables are not necessarily uncorrelated (Tennekes and Lumley, 1972).

7.3.1 Initial Turbulence Developing Region

Jets are created immediately behind the holes in the grid. These jets are intersected by wakes generated behind the solid portion of the grid, resulting in a highly dynamic, anisotropic, and inhomogeneous flow. These jet-wake interactions produce a significant amount of shear, and it is this shear which spawns large turbulent eddies. In this turbulence-producing and developing region, much of the energy is centered around a wavelength on the order of the grid or mesh size (i.e., solid distance between adjacent holes and/or

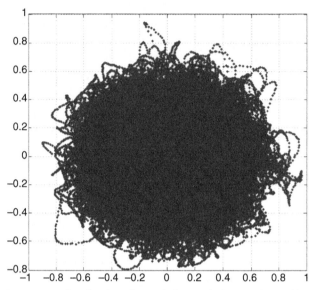

Figure 7.6 Covariance of orificed, perforated plate turbulence. *(Created by R. Liu).*

hole size), which roughly corresponds to the size of the coherent turbulent eddies shed from the solid of the grid. The nonlinear terms in the Navier-Stokes equation then start to redistribute this energy over a broader range of eddy sizes. Depending on certain factors, including the dimensions associated with the grid, this turbulence development region typical lasts until approximately 20 times the characteristic mesh size M associated with the grid.

7.3.2 Power-Law Decay Region

The "fully developed" stage is reached when the kinetic energy is distributed over a wide range of vortical structures (eddies); that is, from the largest scale, which is typically approximated as the energy-containing integral length, down to the smallest dissipative, Kolmogorov microscale. We see that in fully developed high Reynolds number turbulence, the bulk of both the energy and the enstrophy (vorticity) are literally held in two mutually exclusive groups of eddies. The vorticity, which underpins the large eddies via the Biot-Savart law, is weak and dispersed, making little contribution to the net enstrophy. On the other hand, the small eddies are composed of intense patches of vorticity, and so they dominate the enstrophy. Nevertheless, these small eddies make little contribution to the net kinetic energy because they are so small.

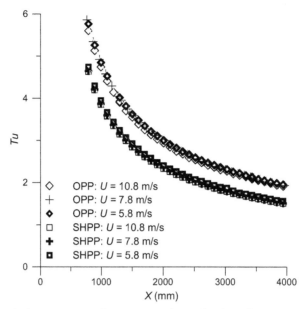

Figure 7.7 Turbulence intensity downstream of an orificed, perforated plate versus that downstream of a straight-hole perforated plate. *(Created by R. Liu).*

This "Power-Law Decay" stage, where the turbulence becomes approximately homogeneous, generally occurs beyond 20 grid or mesh sizes downstream of a typical grid. Figure 7.7 depicts the power-law decay of the fluctuating turbulence downstream of an orificed perforated plate versus that downstream of a straight-hole perforated plate (SHPP) (Liu and Ting, 2007). It is interesting to note the invariant of the decay with respect to the Reynolds number when changing the free-stream velocity from 5.8 to 10.8 m/s. Moreover, the OPP produced roughly 25% higher turbulence intensity than the SHPP.

Davidson (2004) called the turbulence after it reaches the "fully developed" or "asymptotic" stage "freely evolving" or "freely decaying" turbulence. At this stage, there is virtually no interaction between the mean flow, which is more or less uniform, and the turbulence itself. The only function of the mean flow is to carry the turbulence through the tunnel.

In this "decay of fully developed turbulence" phase, unlike the "initial developing" phase, there is significant energy dissipation. This dissipation is mostly carried out via the smallest eddies because their turn-over time, which also turns out to be their break-up time, is much smaller than that of the large eddies. Recall from Chapter 4 that $\eta/u_\eta \ll \Lambda/u$. The dissipation rate per unit mass may be interpreted as

$$d^2_u/dt \sim -u^2/(\Lambda/u) \sim$$
$$- (\text{energy of large eddies})/(\text{turnover time of large eddies}) \qquad (7.4)$$

Incidentally, the time scale of turbulent kinetic energy decay is equal to the characteristic time of the energy-containing eddies (Λ/u). In other words, kinetic energy decay is the process of the destruction of large eddies. Note that viscosity does not appear here, as the rate of dissipation depends on the amount of energy being dissipated, whereas viscosity sets the size of the dissipating eddies by keeping u_η/η on the order of unity; that is, smaller eddies would be nullified by viscosity, while larger ones would be unaffected by it.

In statistically steady (or quasi-steady) turbulence, the rate of kinetic energy destruction at small scales is equal to the energy transfer rate through Richardson's energy cascade, which is controlled by the "break-up" of large eddies. Hence, we have

$$\varepsilon\{= 2\nu S_{ij} S_{ij} \sim \nu u^2_\eta/\eta^2\} \approx \Pi[\sim u^2/(\Lambda/u) = u^3/\Lambda] \qquad (7.5)$$

from which we can relate the small and large scales via

$$\nu u^2_\eta/\eta^2 \sim u^3/\Lambda \qquad (7.6)$$

Recall that the energy cascade process is driven by inertia. Viscosity plays a role only when the eddy size reaches the dissipation scale. In other words, viscosity provides a dustbin for energy at the end of the cascade but does not influence the cascade itself.

Accordingly, most of the information associated with the initial conditions is lost in the process of creating the turbulence. This well-accepted supposition is nevertheless somewhat of a leap of faith. No wonder Batchelor (1953) made the following statement:

"We put our faith in the tendency for dynamical systems with a large number of degrees of freedom, and with coupling between those degrees of freedom, to approach a statistical state which is independent (partially, or wholly) of the initial conditions."

Accepting the premise that turbulence is indeed forgetful, different types of grid may be used to generate turbulence in which the statistical properties of the fully developed turbulence are roughly the same for the corresponding u and Λ. In reality, the turbulence never becomes truly isotropic. Somewhat like elephants, turbulence seems to retain a long-term memory of certain things. This robust information is associated with the dynamical invariants of the flow; that is, it is retained by the turbulence as a direct result of the laws of conservation of linear and angular momentum. For conventional grids, it is found that

$$\overline{v^2} = \overline{w^2} \approx 0.75\overline{u^2} \tag{7.7}$$

This seems to show the remarkable memory of turbulent flows for their initial conditions.

To avoid flow instabilities which cause a non-uniform mean velocity profile, the open area must be approximately 60% or greater. A smaller open area can lead to prevailing non-uniformity in the flow. On the other hand, larger openings are less effective in producing turbulence.

Undeniably, the decay region is the most studied grid turbulence regime. It is thus not surprising to see the increasingly reinvigorated interest in recent years; see Babuin et al., 2014; Isaza et al., 2014; Kitamura et al., 2014; Meldi et al., 2014; Sinhuber et al., 2015; Torrano et al., 2015, and Vassilicos, 2015, among many others. As an introductory textbook, we restrict the coverage to only the basics of well-established, canonical research on decaying grid turbulence. Before we scrutinize this decaying regime further, let us complete the introduction of the two remaining grid turbulence regimes.

7.3.3 Dominating Large-Scale Region

The dominating large-scale region is typically not considered unto itself, as it is simply the area between the power-law decay region and the final decay region. Nevertheless, it is worth noting that some researchers have proposed different ways of sorting out flow regions; for example, Skrbek et al. (2000) identified four regimes of decaying grid turbulence with the help of helium II.

7.3.4 Final Decay Region

Further downstream, the faster decaying smaller eddies are gone, leaving the slower decaying large scales. Because they decay slowly, there is a notable lack of interaction between the large eddies. It is obvious that the corresponding Reynolds number is sufficiently small as the flow enters this region. Problem 7.4 suggests a Reynolds number on the order of 10 as a possible threshold.

7.4 DECAY OF HOMOGENEOUS ISOTROPIC TURBULENCE

Assume that we can approximate grid turbulence by the isotropic assumption. For isotropic turbulence, two times total kinetic energy per unit mass is

$$\overline{q^2} = \overline{u^2} + \overline{v^2} + \overline{w^2} = \overline{3u^2} \tag{7.8}$$

We should underscore the fact that rigorously speaking, real grid turbulence is neither isotropic nor homogeneous (Ertunç et al., 2010). For typical grids, the actual measured grid turbulence is

$$\overline{q^2} = \overline{u^2} + \overline{v^2} + \overline{w^2} = \overline{2.5u^2} \tag{7.9}$$

A properly designed OPP with 43% solidity produces a significantly more isotropic turbulence with $\overline{u^2} \approx 1.1\overline{v^2}$ or $\overline{q^2} \approx 2.8\overline{u^2}$ (Liu et al., 2007).

We will proceed to derive expressions of the variations of key turbulence characteristics with respect to distance downstream of a grid. Invoking the ergodic hypothesis as verified by Djenidi et al. (2013), these expressions are also applicable when considered with respect to time. The general layout in Wilson (1989) is emulated. First, the decrease in turbulence intensity with streamwise distance is derived. This is followed by the Taylor microscale, dissipation rate, and integral length in the next section. The streamwise alterations of the relative magnitude among the three primary length scales (integral length, Taylor microscale, and Kolmogorov scale) are formulated in the subsequent section.

For turbulence that is homogeneous in the y- and z-directions and varies slowly in the x-direction, we may assume locally homogeneous turbulence. Then all transport terms vanish in the kinetic energy equation, and we are left with

$$\frac{1}{2}\frac{d\overline{q^2}}{dt} = P_{\text{turb}} - \varepsilon \tag{7.10}$$

Because the mean flow is homogeneous

$$\overline{V} = \overline{W} = 0 \tag{7.11}$$

and

$$\partial \overline{U} / \partial x_i = 0 \tag{7.12}$$

therefore

$$P_{\text{turb}} = 0 \tag{7.13}$$

that is, there is no turbulence production, except close to the grid.

Assuming isotropic dissipation, Taylor (1935) showed that

$$\varepsilon = 15\nu \overline{\left(\frac{\partial u}{\partial x}\right)^2} \tag{7.14}$$

from which the microscale $\lambda_g = \lambda_f/2$ and from the definition of λ_f

$$\overline{\left(\frac{\partial u}{\partial x}\right)^2} \equiv \frac{\overline{u^2}}{\lambda_g^2} \tag{7.15}$$

so that

$$\varepsilon = 15\nu \frac{\overline{u^2}}{\lambda_g^2} \tag{7.16}$$

Substituting Eqs (7.8 and 7.16) into the energy balance equation, Eq. (7.10) gives

$$\frac{3}{2}\frac{d\,\overline{u^2}}{dt} = -15\nu \frac{\overline{u^2}}{\lambda_g^2} \tag{7.17}$$

Assume a power law function for the decay of $\overline{u^2}$, with a constant C_1 and an exponent n

$$\frac{\overline{u^2}}{\overline{U}^2} = C_1 \left(\frac{x}{M} - \frac{x_0}{M}\right)^n \tag{7.18}$$

or, since $x = \overline{U}t$ according to Taylor's frozen hypothesis, we have

$$\frac{\overline{u^2}}{\overline{U}^2} = C_1 \left(\frac{\overline{U}}{M}\right)^n (t - t_0)^n \tag{7.19}$$

where x_0 is the distance from the grid required to generate the turbulence in the far wake. Depending largely on the grid, x_0/M can be up to 30–40. For $x/M < x_0/M$, the turbulent flow is inhomogeneous, anisotropic and $P_{turb} > 0$.

Then, substitute Eq. (7.19) into Eq. (7.17) to get

$$n\left(\frac{C_1\overline{U}^{2+n}}{M^n}\right)(t - t_0)^{n-1} = -10\frac{\nu}{\lambda_g^2}\left(\frac{C_1\overline{U}^{2+n}}{M^n}\right)(t - t_0)^n \tag{7.20}$$

This can be reduced into

$$\lambda_g^2 = -10\frac{v}{n}\left(t - t_0\right)$$ (7.21)

We see that λ_g is proportional to $\sqrt{t}$ and that as time goes on, or as we move farther downstream, Taylor microscale increases in size. Eventually, larger eddies are forced to provide the dissipation ε as the supply of small eddies is used up.

If we insert λ_g^2 from Eq. (7.21) and $\overline{u^2}$ from Eq. (7.19) into Eq. (7.16) for dissipation, we get

$$\varepsilon = 15v\frac{\left(\dfrac{C_1\overline{U}^{2+n}}{M^n}\right)\left(t - t_0\right)^n}{-10\dfrac{v}{n}\left(t - t_0\right)} = -\frac{3n}{2}\left(\frac{C_1\overline{U}^{2+n}}{M^n}\right)\left(t - t_0\right)^{n-1}$$ (7.22)

or

$$\varepsilon = -\frac{3nC_1}{2}\left(\frac{\overline{U}^3}{M}\right)\left(\frac{x}{M} - \frac{x_0}{M}\right)^{n-1}$$ (7.23)

This shows clearly that a change in viscosity v will only affect the scale of dissipation λ_g and not the dissipation ε itself. The dissipation is controlled only by the large-scale motion $\overline{U}^3/M$, which is independent of viscosity.

7.5 ESTIMATING THE INTEGRAL SCALE VARIATION

As expounded upon in Chapter 4, dissipation is a passive process, which depends on the amount of energy being passed down through the energy cascade from the large eddies; in other words, dissipation is controlled directly by large eddy motions. As such, we can deduce the large scale once we know the rate of turbulent kinetic dissipation, that is, the decay rate in the absence of production as per Eq. (7.10). Specifically

$$\varepsilon \approx B_0\frac{u_t^3}{l}$$ (7.24)

where the equivalent three-dimensional fluctuating velocity

$$u_t^2 = \frac{\overline{q^2}}{3} = \frac{\overline{u^2} + \overline{v^2} + \overline{w^2}}{3}$$ (7.25)

It is obvious that $u_t^2 = \overline{u^2}$ in isotropic turbulence. Hinze (1975) stated Eq. (7.24) and then used the analysis of frequency spectra to show that

$$l = \Lambda_f / 0.75 \tag{7.26}$$

Using the isotropic relation

$$\Lambda_f = 2\Lambda_g \tag{7.27}$$

where Λ_f is the streamwise correlation scale and Λ_g is the cross-stream correlation scale, we find that

$$l = 2.66\,\Lambda_g \tag{7.28}$$

Also, Hinze (1975) used experimental frequency spectra to obtain the proportionality constant

$$B_0 \approx 0.8 \tag{7.29}$$

Using an orificed perforated plate, on the other hand, Liu and Ting (2007) obtained a value of 1.08 for B_0, which is much closer to unity. With $B_0 \approx 0.8$ and $l = 2.66\,\Lambda_g$, we can thus rewrite Eq. (7.24) as

$$\varepsilon \approx 0.30 \frac{\left(\overline{u^2}\right)^{3/2}}{\Lambda_g} \tag{7.30}$$

All three parameters involved are fairly well-defined, as we have already seen in Chapter 4. Using Eq. (7.19) for $\overline{u^2}$ and equating Eqs (7.30 and 7.23), we have

$$\frac{0.30}{\Lambda_g}\left(\frac{C_1 U^{2+n}}{M^n}\right)^{3/2}\left(t-t_0\right)^{3n/2} = -\frac{3n}{2}\left(\frac{C_1 U^{2+n}}{M^n}\right)\left(t-t_0\right)^{n-1} \tag{7.31}$$

This can be manipulated to give the integral length as a function of time or distance downstream of the grid

$$\Lambda_g = \frac{0.6}{3n}\left(\frac{C_1 U^{2+n}}{M^n}\right)^{1/2}\left(t-t_0\right)^{\frac{2+n}{2}} \tag{7.32}$$

Dividing this by the mesh width M and using the relation $x = \overline{U}t$, we have the variation of the integral length with respect to grid mesh size and distance downstream of the grid

$$\frac{\Lambda_g}{M} = -0.2\frac{C_1^{1/2}}{n}\left(\frac{x}{M} - \frac{x_0}{M}\right)^{\frac{n+2}{2}}$$ (7.33)

The $\Lambda_g \propto C_1^{1/2}$ relation, that is, the size of integral scale, is proportional to the square root of the turbulence decay coefficient, as has been verified experimentally by Liu and Ting (2007).

Early estimates by investigators such as Batchelor and Townsend (1948a) suggested that $n = -1.0$. Notwithstanding this, Hinze (1975) presented an analysis using correlation invariants that showed n as equal to $-6/5$ for Saffman's invariant or $-10/7$ for Loitsiansky's "invariant." However, Synge and Lin's (1943) experimental data appear to indicate that the Loitsiansky invariant does not exist; see Pullin and Saffman (1998), for example. Comte-Bellot and Corrsin (1966) used a contraction to eliminate grid turbulence, forcing it to be more isotropic. They found $n = -1.28$, which is in fairly good agreement with Hinze's analysis using Saffman's invariant. Liu and Ting (2007) found $n = -1.15$ for both orificed, perforated plates and straight-hole (finite thickness) perforated plates. With many veridical experiments emanating values of n in the vicinity of Saffman's invariant, it is not surprising that experts have wondered if grid turbulence is indeed Saffman turbulence (Krogstad and Davidson, 2010).

7.6 KOLMOGOROV SCALE IN DECAYING GRID TURBULENCE

As the dissipation rate ε changes with distance from the grid, so will the Kolmogorov length η because the Kolmogorov length is defined as

$$\eta \equiv \left(\frac{v^3}{\varepsilon}\right)^{1/4}$$ (7.34)

This can be recast as

$$\varepsilon = \frac{v^3}{\eta^4}$$ (7.35)

Equating this to isotropic dissipation, Eq. (7.14), we have

$$\frac{\lambda_g}{\eta} = 2.0\left(\frac{\overline{u^2}\lambda_g^2}{v^2}\right)^{1/4}$$ (7.36)

From Eq. (7.18), we see

$$\overline{u^2} = C_1 \overline{U}^2 \left(\frac{x}{M} - \frac{x_0}{M} \right)^n \tag{7.37}$$

Since $\overline{U} = x/t$ and hence, $t = x/\overline{U}$, we can re-express Eq. (7.21) as

$$\lambda_g^2 = -10 \frac{\nu}{n} M \left(\frac{x}{M\overline{U}} - \frac{x_0}{M\overline{U}} \right) \tag{7.38}$$

Substituting Eqs (7.37 and 7.38) into Eq. (7.36), we get

$$\frac{\lambda_g}{\eta} = 2.0 \left[\frac{C_1 \overline{U}^2}{\nu^2} \left(\frac{x}{M} - \frac{x_0}{M} \right)^n \left(\frac{-10\nu M}{n\overline{U}} \right) \left(\frac{x}{M} - \frac{x_0}{M} \right) \right]^{1/4} \tag{7.39}$$

This can be simplified, using the Reynolds number based on the mesh, $\text{Re}_M \equiv \overline{U}M/\nu$, into

$$\frac{\lambda_g}{\eta} = \left(\frac{150 C_1}{-n} \right)^{1/4} \text{Re}_M^{1/4} \left(\frac{x}{M} - \frac{x_0}{M} \right)^{\frac{n+1}{4}} \tag{7.40}$$

We see that this Taylor microscale-Kolmogorov scale ratio varies relatively slowly as the turbulence decays downstream of the grid. Notably, if we invoke Saffman's invariant, $n = -1.2$, we have

$$\frac{\lambda_g}{\eta} \propto \left(\frac{x}{M} - \frac{x_0}{M} \right)^{-0.05} \tag{7.41}$$

This is discernibly smaller compared to the "swelling" in integral scale, as the turbulence and the smaller eddies decay, Eq. (7.33)

$$\frac{\Lambda_g}{M} \propto \left(\frac{x}{M} - \frac{x_0}{M} \right)^{0.40} \tag{7.42}$$

The slower change in λ_g/η is no surprise, as both Taylor microscales and Kolmogorov scales portray the small, dissipating eddies. The decay in the fluctuating intensity from Eq. (7.18)

$$\frac{\sqrt{\overline{u^2}}}{\overline{U}} \propto \left(\frac{x}{M} - \frac{x_0}{M} \right)^{-0.6} \tag{7.43}$$

is relatively faster.

To obtain the relative variation of the Taylor microscale with respect to integral length of distance downstream of the grid, we combine Eqs (7.21 and 7.33) (or simply Eqs (7.38 and 7.42) to get

$$\frac{\lambda_g}{\Lambda_g} \propto \left(\frac{x}{M} - \frac{x_0}{M} \right)^{0.1} \tag{7.44}$$

We see that the change in this small-large scale ratio is significantly faster than the smallest small-scale ratio described by Eq. (7.41). As the turbulence decays downstream, the energy cascade is progressively shortened at the more dissipative, faster decaying, small-scale end. Consequently, the respective small and large scales of the remaining energy cascade have to be continuously reassigned. This leads to quickly increasing Kolmogorov scale followed immediately by Taylor microscale, while the integral length changes relatively slower.

Furthermore, the following relationships can be seen from the conventional grid turbulence plots in Hinze (1975)

$$\lambda_g^2 \propto x \Rightarrow \lambda_g \propto \sqrt{t} \tag{7.45}$$

In other words, Taylor microscale increases with the square root of time or distance downstream of the grid. This has been inferred from Eq. (7.21). On the other hand, the turbulence intensity decreases much faster

$$\left(\sqrt{\overline{u^2}} \right)^2 \propto \left(x - x_0 \right)^{-1.2} \tag{7.46}$$

Some data support a power of -1.0 instead of -1.2. Most conventional grids lead to quite anisotropic turbulence when cross-stream intensities are only 65% of the streamwise value; that is

$$\overline{v^2} = \overline{w^2} \approx 0.65 \overline{u^2} \tag{7.47}$$

7.7 SPECTRAL SPACE

There are some advantages to recasting the equations of turbulence from real space (defined as space $\mathbf{r}$ and time t) into Fourier space (defined as wave-number space $\mathbf{k}$ and time t). Fourier transform acts like a filter, sorting out or differentiating the different scales present within a fluctuating signal.

Turbulence measurements are typically for Eulerian spectra, where fluctuations of randomly oriented eddies are quantified as they are swept by a probe at a velocity U. Consequently, only a one-dimensional slice of the three-dimensional spectrum is measured. The spectral density with frequency n is $E_1(n)$ for the velocity component u. The fraction of $\overline{u^2}$ between n and $n + dn$ is $E_1(n)dn$ and

$$\overline{u^2} = \int_0^\infty E_1(n)\,dn \qquad (7.48)$$

One problem that arises with expressing spectra in terms of frequency is the false increase in frequency when the mean velocity U is increased. In other words, Fourier transforming a higher velocity gives a higher frequency, but there may or may not be a corresponding increase in the turbulent fluctuation frequencies. As we are interested in frequencies which are directly related to the eddy sizes and not the artifact of changing frequencies associated with variations in sweeping velocity, it is better to express frequency in terms of the wave-number k_1, where

$$k_1 \equiv 2\pi n/U = 2\pi/\text{wavelength} \qquad (7.49)$$

where subscript "1" is used to distinguish the one-dimensional wave-number from the three-dimensional wave-number k.

We see that

$$E_1(k_1)\,dk_1 = E_1(n)\,dn \qquad (7.50)$$

But $dk_1 = 2\pi dn/U$, and hence

$$E_1(k_1) = E_1(n)U/(2\pi) \qquad (7.51)$$

Therefore

$$\overline{u^2} = \int_0^\infty E_1(k_1)\,dk_1 \qquad (7.52)$$

Figure 7.8 is a schematic of a spectral space. The dashed line indicates the effects of Reynolds number on the frozen turbulence assumption; whereas for the smallest eddies, some amount of dissipation is expected, especially at higher Re.

As discussed in Chapter 4, at sufficiently high Reynolds number, there exists an inertial sub-range in the turbulence spectrum, which can be represented by a simple power function of the form

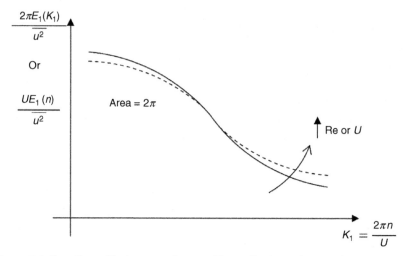

Figure 7.8 The effect of Re in spectral space. *(Created by A. Vasel-Be-Hagh).*

$$E_1(k_1) = A_1\, \varepsilon^a\, k_1^b \tag{7.53}$$

where exponents "*a*" and "*b*" can be found via dimensional analysis. Note that $k_1 \equiv 2\pi n/U$ has units of m^{-1} and from Eq. (7.51), the spectral density

$$E_1(k_1) \sim [m^3/s^2] \tag{7.54}$$

With $\varepsilon \sim [s^2/m^3]$, we have for Eq. (7.53)

$$[m^3/s^2] = [m^2/s^3]^a\,[1/m]^b \tag{7.55}$$

which yields $a = 2/3$ and $b = -5/3$, or

$$E_1(k_1) = A_1\, \varepsilon^{2/3}\, k_1^{-5/3} \tag{7.56}$$

Hinze (1975) found that for isotropic turbulence, $A_1 \approx 0.56$ for the x component of the turbulence velocity.

Note that the spectrum of the cross-stream components are not the same as $E_1(k_1)$ of the streamwise component, even in isotropic turbulence. The differences are caused by "aliasing" of the three-dimensional spectrum $E(k)$ of $\overline{q^2}/2$ by the one-dimensional slice taken by a sensor that sees the wave field swept at speed U. According to Hinze (1975), for isotropic turbulence

$$E_2(k_1) = E_3(k_1) = \tfrac{1}{2}[E_1(k_1) - k_2\,\partial E_1(k_1)/\partial k_1)] \tag{7.57}$$

In the inertial sub-range, we have

$$E_2(k_1) = E_3(k_1) = 4\,A_1\,\varepsilon^{2/3}\,k_1^{-5/3}/3 \tag{7.58}$$

for isotropic turbulence.

Figure 7.9 is a spectra plot of the OPP turbulence at 10.8 m/s free-stream velocity. We see that the three spectra corresponding to 20, 60, and 100 hole diameters downstream of the OPP collapse nicely unto each other, which is not the case for the SHPP turbulence (see Liu and Ting [2007]). For $k_1\Lambda$ of less than unity, which corresponds to the largest structures, a slight departure from the self-preservation state is noted. This is somewhat expected, as the largest structures are expected to carry some "genetic biases" from the jetwake interactions immediately behind the OPP. Some very high frequency noise is also noted.

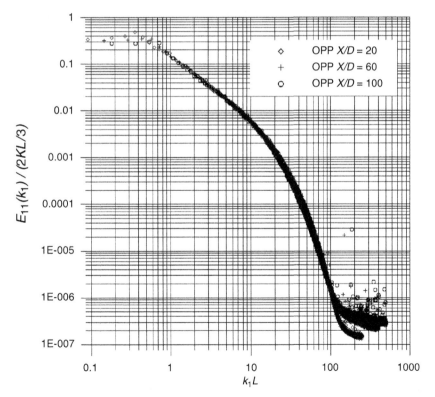

Figure 7.9 Normalized streamwise turbulence velocity spectrum $E_1/(2k\Lambda/3)$ at $U =$ 10.8 m/s downstream of the OPP. *(Created by R. Liu).*

Problems

Problem 7.1. The decay of turbulence
The decay of turbulent kinetic energy is typically expressed in the power law form,

$$kE/\text{mass} = A \,(t - t_0)^{-n}.$$

1. Prove that the smallest value of n is unity, i.e., $n \geq 1$.
2. How does n change as the turbulence enters into the "final" period of decay?

Problem 7.2. The anisotropy of grid turbulence
Carry out a refined analysis of grid turbulence that accounts for the nonisotropic nature of the turbulence. Assume that $\overline{v^2} = \overline{w^2} \approx 0.75\overline{u^2}$ and derive the power-law decay functions for the Taylor microscale λ_g and the integral Λ_g. Express the equations in terms of $\overline{u^2}$ rather than $\overline{q^2}$. Discuss differences between this and isotropic results.

Problem 7.3. Integral-Taylor scale ratio in isotropic grid turbulence
Using the usual isotropic results, derive a relationship for the ratio of macro to microscale Λ_g / λ_g as a function of distance x/M. How is the ratio affected by $Re = UM/v$, where M is the mesh size? Compare this to general scaling results.

Problem 7.4. Initial versus final turbulence decay
A cubical box of volume L^3 filled with fluid is shaken to generate a sufficient amount of turbulence and then the turbulence is left to decay.

1. Derive an expression for the decay of the kinetic energy $3u^2/2$ as a function of time.
2. When the turbulence decays to Re ($= uL/v$) of less than 10, the inviscid estimate $\varepsilon = u^3/L$ may be replaced by an estimate of the type $\varepsilon = cvu^2/L^2$, because the weak eddies remaining at low Re lose their energy directly to viscous dissipation. Compute c by requiring that the dissipation rate is continuous at $uL/v = 10$.
3. Derive an expression for the decay of the kinetic energy during the final decay period when $uL/v < 10$.
4. If $L = 1$ m, $v = 1.5 \times 10^{-7}$ m^2/s and $u = 1$ m/s at time $t = 0$, how long does it take before the turbulence enters the final period of decay?

REFERENCES

Babuin, S., Varga, E., Skrbek, L., 2014. The decay of forced turbulent coflow of He II past a grid. J. Low Temp. Phys. 175, 324–330.

Batchelor, G.K., 1953. The Theory of Homogeneous Turbulence. Cambridge University Press, Cambridge.

Batchelor, G.K., Townsend, A.A., 1947. Decay of vorticity in isotropic turbulence. Proc. R. Soc. Lond. Ser. A 190 (1023), 534–550.

Batchelor, G.K., Townsend, A.A., 1948a. Decay of isotropic turbulence in the initial period. Proc. R. Soc. Lond. Ser. A 193 (1035), 539–558.

Batchelor, G.K., Townsend, A.A., 1948b. Decay of isotropic turbulence in the final period. Proc. R. Soc. Lond. Ser. A 194 (1039), 527–543.

Comte-Bellot, G., Corrsin, S., 1966. The use of a contraction to improve the isotropy of grid-generated turbulence. J. Fluid Mech. 25 (4), 657–682.

Davidson, P.A., 2004. Turbulence: An Introduction for Scientists and Engineers. Oxford University Press, Oxford.

Djenidi, L., Tardu, S.F., Antonia, R.A., 2013. Relationship between temporal and spatial averages in grid turbulence. J. Fluid Mech. 730, 593–606.

Ertunç, Ö., Özyilmaz, N., Lienhart, H., Durst, F., Beronov, K., 2010. Homogeneity of turbulence generated by static-grid structures. J. Fluid Mech. 654, 473–500.

Isaza, J., Salazar, R., Warhaft, Z., 2014. On grid-generated turbulence in the near- and far field regions. J. Fluid Mech. 753, 402–426.

Hinze, J.O., 1975. Turbulence, second ed. McGraw-Hill, USA.

Kitamura, T., Nagata, K., Sakai, Y., Sasoh, A., Terashima, O., Saito, H., Harasaki, T., 2014. On invariants in grid turbulence at moderate Reynolds numbers. J. Fluid Mech. 738, 378–406.

Krogstad, P.Å., Davidson, P.A., 2010. Is grid turbulence Saffman turbulence? J. Fluid Mech. 642, 373–394.

Lin, C.C., 1949. Note on the law of decay of isotropic turbulence. Proc. Natl. Acad. Sci. U.S.A. 34, 540–543, 1948.

Liu, R., Ting, D.S-K., 2007. Turbulent flow downstream of a perforated plate: sharp-edged orifice versus finite-thickness holes. J. Fluids Eng. 129, 1164–1171.

Liu, R., Ting, D.S-K., Checkel, M.D., 2007. Constant Reynolds number turbulence downstream of an orificed, perforated plate. Exp. Therm Fluid Sci. 31, 897–908.

Liu, R., Ting, D.S-K., Rankin, G.W., 2004. On the generation of turbulence with a perforated plate. Exp. Therm. Fluid Sci. 28, 307–316.

Meldi, M., Lejemble, H., Sagaut, P., 2014. On the emergence of non-classical decay regimes in multiscale/fractal generated isotropic turbulence. J. Fluid Mech. 756, 816–843.

Mi, J., Nathan, G.J., Nobes, D.S., 2001. Mixing characteristics of axisymmetric free jets from a contoured nozzle, an orifice plate and a pipe. J. Fluids Eng. 123 (4), 878–883.

Portfors, E.A., Keffer, J.F., 1969. Isotropy in initial period grid turbulence. Phys. Fluids A 12, 1519–1521.

Pullin, D.I., Saffman, P.G., 1998. Vortex dynamics in turbulence. Annu. Rev. Fluid Mech. 30, 31–51.

Simmons, L.F.G., Salter, C., 1934. Experimental investigation and analysis of the velocity variations in turbulent flow. Proc. R. Soc. Lond. Ser. A 145, 215–253.

Skrbek, L., Niemela, J.J., Donnelly, R.J., 2000. Four regimes of decaying grid turbulence in a finite channel. Phys. Rev. Lett. 85 (14), 2973–2976.

Sinhuber, M., Bodenschatz, E., Bewley, G.P., 2015. Decay of turbulence at high Reynolds numbers. Phys. Rev. Lett. 114 (3)/034501(5).

Stewart, R.W., Townsend, A.A., 1951. Similarity and self-preservation in isotropic turbulence. Philos. Trans. R. Soc. Lond. Ser. A 243 (867), 359–386.

Synge, J.L., Lin, C.C., 1943. On a statistical model of isotropic turbulence. T. Roy. Soc. Can. 37, 45–79.

Taylor, G.I., 1935. Statistical theory of turbulence II: measurements of correlation in the Eulerian representation of turbulent flow. Proc. R. Soc. Lond. Ser. A 151 (873), 444–454.

Tennekes, H., Lumley, J.L., 1972. A First Course in Turbulence. MIT Press, Cambridge.

Torrano, I., Tutar, M., Martinez-Agirre, M., Rouquier, A., Mordant, N., Bourgoin, M., 2015. Comparison of experimental and RANS-based numerical studies of the decay of grid-generated turbulence. J. Fluids Eng. 137/061203, 1–12.

Tresso, R., Munoz, D.R., 2000. Homogeneous, isotropic flow in grid generated turbulence. J. Fluids Eng. 122, 51–56.

Vassilicos, J.C., 2015. Dissipation in turbulent flows. Ann. Rev. Fluid Mech. 47, 95–114.

Von Kármán, T., Lin, C.C., 1949. On the concept of similarity in the theory of isotropic turbulence. Rev. Mod. Phys. 21, 516–519.

Wilson, D.J., 1989. Mec E 632 Turbulent Fluid Dynamics. University of Alberta, Edmonton.

CHAPTER 8

Vortex Dynamics

The scientist does not study nature because it is useful; he studies it because he delights in it, and he delights in it because it is beautiful. If nature were not beautiful, it would not be worth knowing, and if nature were not worth knowing, life would not be worth living.

–Henri Poincaré

Contents

Chapter Objectives

- To comprehend the basics of vortex dynamics.
- To appreciate the importance of vortex dynamics in flow turbulence.
- To describe flow turbulence in terms of simple vortices.
- To model rapid changes in turbulence using rapid distortion theory.

NOMENCLATURE

A	Area
I	Moment of inertia
L	Length
m	Mass
P	Pressure
r	Radius
s	Path, distance
T	Temperature
u	The fluctuating component of the velocity (in the x direction)
U	Time-averaged velocity (in the x direction)

Basics of Engineering Turbulence
http://dx.doi.org/10.1016/B978-0-12-803970-0.00008-8

v	The fluctuating component of the velocity in the y direction
V	Time-averaged velocity in the y direction
V_n	Normal velocity
V_t	Tangential velocity
w	The fluctuating component of the velocity in the z direction
W	Time-averaged velocity in the z direction
x, y, z	Cartesian coordinates

Greek Symbols

α	Angle
β	Angle
Γ	Circulation
γ	Specific heat ratio
θ	Angle
μ	Dynamic (absolute) viscosity
ν	Kinematic viscosity
ρ	Density
φ	A scalar
Ω	Angular speed
ω	Vorticity

8.1 INTRODUCTION

One salient characteristic of flow turbulence is that it is highly vortical. For well-developed turbulence, there also exists an "energy ladder" for conveying the turbulent kinetic energy down a cascade of eddying motions, which are decreasing size, but increasing in number. By its very nature, most of the kinetic energy is associated with the large eddies, while the smallest eddies contain most of the vorticity. We learned from Chapter 4 that the highly vortical, small eddies are directly related to the large, energy containing eddies, with intermediate-sized eddies acting as a passive energy transfer passage. As such, a sound comprehension of the flow turbulence requires a good understanding of the vortical structures. In other words, the study of vortex dynamics appears to be an appropriate means for understanding and describing turbulence; see Pullin and Saffman (1998), Bernard (2013). Some basic definitions, equations, and theories will first be introduced before applying three orthogonal vortices to interpret some elements of flow turbulence.

A vortex is simply the rotating motion of a multitude of material particles around a common center. The paths of the individual particles do not have to be circular; that is, they may be asymmetrical, as portrayed in Fig. 8.1. Furthermore, a vortex can be two-dimensional such as those

(a) 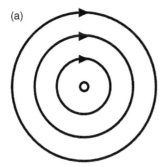   (b)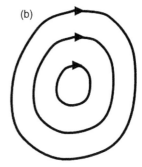

Figure 8.1 Two-dimensional vortices: (a) concentric circular vortex, (b) asymmetrical vortex. *(Created by B. Cheung).*

illustrated in Fig. 8.1, or it can be three-dimensional as depicted in Fig. 8.2. Recall that an important feature of flow turbulence is three-dimensionality. As such, flow turbulence can be perceived as a myriad of interacting three-dimensional vortices.

8.2 VORTICITY

The angular velocity of matter at a point in continuum space is called vorticity. While there is no vortex without vorticity, a vorticity field does not have to represent a vortex. Vorticity can be defined as (Saffman, 1992)

$$\vec{\omega} = curl\vec{V} = \nabla \times \vec{V} = e_{ijk}\hat{e}_i \frac{\partial V_k}{\partial x_j} = \begin{vmatrix} \hat{e}_1 & \hat{e}_2 & \hat{e}_3 \\ \dfrac{\partial}{\partial x_1} & \dfrac{\partial}{\partial x_2} & \dfrac{\partial}{\partial x_3} \\ V_1 & V_2 & V_3 \end{vmatrix} \tag{8.1}$$

In terms of $x, y, z, U, V,$ and W, we have

$$\vec{\omega} = \begin{vmatrix} \hat{i} & \hat{j} & \hat{k} \\ \dfrac{\partial}{\partial x} & \dfrac{\partial}{\partial y} & \dfrac{\partial}{\partial z} \\ U & V & W \end{vmatrix} \tag{8.2}$$

This can be expanded to give

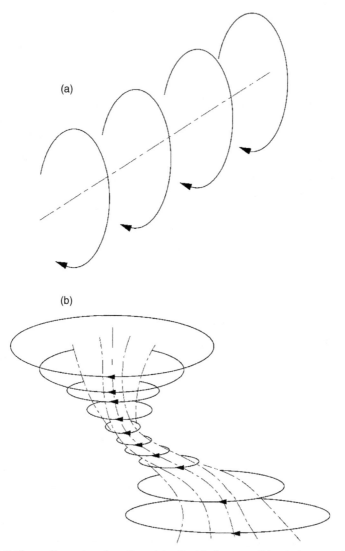

(a)

(b)

Figure 8.2 Three-dimensional vortices: (a) cylindrical vortex, (b) spiral vortex. *(Created by N. Cao).*

$$\vec{\omega} = \left(\frac{\partial W}{\partial y} - \frac{\partial V}{\partial z} \right)\hat{i} + \left(\frac{\partial U}{\partial z} - \frac{\partial W}{\partial x} \right)\hat{j} + \left(\frac{\partial V}{\partial x} - \frac{\partial U}{\partial y} \right)\hat{k} \qquad (8.3)$$

Consider two infinitesimal fluid lines AB and BC, as shown in Fig. 8.3. From time t to $t + dt$, where dt is an infinitesimal time step, these two lines undergo both translation and rotation. The rotation is caused by the

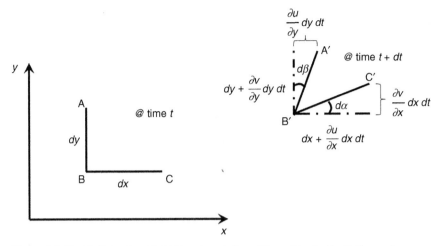

Figure 8.3 Two infinitesimal fluid lines in rotational flow. *(Created by B. Cheung).*

corresponding velocity gradients in the two directions, and it contributes to the vorticity. We can define the angular speed in the z direction, Ω_z as the average rotation rate of the two lines (Currie, 1974; White, 2006; Pritchard and Mitchell, 2015); that is

$$\frac{1}{2}\left(\frac{\partial\alpha}{\partial t} - \frac{\partial\beta}{\partial t}\right) \tag{8.4}$$

For an infinitely small time step

$$d\alpha = \lim_{dt\to 0}\tan^{-1}\left[\frac{\dfrac{\partial V}{\partial x}dx\,dt}{dx + \dfrac{\partial U}{\partial x}dx\,dt}\right] \tag{8.5}$$

becomes

$$d\alpha \approx \lim_{dt\to 0}\tan^{-1}\left(\frac{\partial V}{\partial x}dt\right) = \frac{\partial V}{\partial x}dt \tag{8.6}$$

This can be recast into

$$\frac{d\alpha}{dt} = \frac{\partial V}{\partial x} \tag{8.7}$$

Similarly

$$d\beta = \lim_{dt \to 0} \tan^{-1}\left[\frac{\dfrac{\partial U}{\partial y}dy\,dt}{dy + \dfrac{\partial V}{\partial y}dy\,dt}\right] \approx \lim_{dt \to 0} \tan^{-1}\left(\frac{\partial U}{\partial y}dt\right) = \frac{\partial U}{\partial y}dt \quad (8.8)$$

From which we can obtain

$$\frac{d\beta}{dt} = \frac{\partial U}{\partial y} \quad (8.9)$$

Substituting Eqs 8.7 and 8.9 into Eq. 8.4, we get

$$\Omega_z = \frac{1}{2}\left(\frac{\partial \alpha}{\partial t} - \frac{\partial \beta}{\partial t}\right) = \frac{1}{2}\left(\frac{\partial V}{\partial x} - \frac{\partial U}{\partial y}\right) = \frac{1}{2}\omega_z \quad (8.10)$$

We see that the rotational speed in the z direction is equal to one-half the corresponding vorticity. The above procedure can be repeated for the x and y components to acquire similar expressions for Ω_x and Ω_y. In short, the vorticity is two times the angular velocity

$$\vec{\omega} = \nabla \times \vec{V} = 2\vec{\Omega} \quad (8.11)$$

The difference between a rotational flow and an irrotational flow can be straightforwardly depicted by Fig. 8.4. The fluid in Fig. 8.4a acts like a solid body in a container on a rotating turntable. The changing orientation of label L plainly exhibits that the flow is rotational, specifically, vorticity is everywhere in the tank. When draining the water in a bathtub or a sink, the flow is largely irrotational. This is demonstrated in Fig. 8.4b with zero vorticity everywhere except at the center, which is the point of singularity. In other words, label L does not rotate when placed anywhere except right at the center. As such the corresponding angular velocity U_θ is literally zero everywhere and goes to infinity as we approach the singularity point. Circulation Γ will be explained in the next section.

8.3 KELVIN'S CIRCULATION THEOREM

Among his many admirable contributions, Lord Kelvin (Sir William Thomson) published a series of spearheading treatises on vortex (Kelvin, 1867a, b, 1880). These advancements sprouted from the vortex groundwork

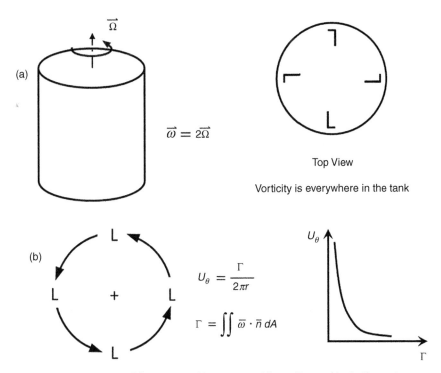

$$\bar{\omega} = 2\bar{\Omega}$$

Top View

Vorticity is everywhere in the tank

$$U_\theta = \frac{\Gamma}{2\pi r}$$

$$\Gamma = \iint \bar{\omega} \cdot \bar{n} \, dA$$

Figure 8.4 (a) Rotational flow versus (b) irrotational flow. *(Created by B. Cheung).*

established by Hermann von Helmholtz (1858). There are also some variations of Helmholtz's theorems regarding a vortex in an inviscid fluid (Prandtl and Tietjens, 1934; Tokaty, 1971; Saffman, 1992; Kundu et al., 2015; Wikipedia, 2015). Let us introduce the concept of vortex lines and vortex tubes before moving further. A vortex line is a line that is everywhere tangent to the vorticity vector, while a tube made of vortex lines is a vortex tube (Wilcox, 2007). Helmholtz's theorems may be expressed in the following three statements: (1) The strength of a vortex tube (its circulation) is constant along its length and with respect to time; (2) A vortex tube cannot end within the fluid, as it must extend to the boundaries of the fluid or form a closed path; (3) Vortex lines move with the fluid. In real fluids, viscosity is finite and hence, vortices decay.

Circulation is defined as the integral of scalar (dot) product of vector velocity times vector displacement around a closed curve at some instant, that is

$$\Gamma \equiv \oint \vec{V} \cdot d\vec{s} = \oint V_t \, ds \qquad (8.12)$$

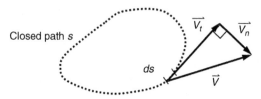

Figure 8.5 Circulation. *(Created by B. Cheung).*

where s is the closed path and V_t is the tangential component of velocity vector $\vec{V}$. This is illustrated in Fig. 8.5, where V_n is the normal component of velocity vector $\vec{V}$.

According to Kelvin (1868), the circulation Γ around any loop in an inviscid fluid remains constant for material lines moving with the fluid. Kelvin's theorem is usually stated for an incompressible fluid, but Batchelor (1967) pointed out that it may be applied to a barotropic compressible flow, which has density that is solely a function of pressure, that is, $\rho = \rho(P)$, rather than $\rho = \rho(P, T)$. All polytropic processes ($P\,\rho^\gamma$ = constant) of an ideal gas satisfy the requirement of barotropic flow, noting that isentropic (constant entropy) and isothermal processes are barotropic.

Moreover, Batchelor (1967) noted that Kelvin's circulation theorem may be used to prove Helmholtz (1858) theorem that vortex tubes move with the fluid by putting a closed loop on the surface of the vortex tube; that is

$$\Gamma_A = \phi V_t\, ds = 2\pi\, r\,(r\Omega) = \text{constant} \tag{8.13}$$

See Fig. 8.6 for more on Eq. (8.13). The only way for Γ_A to always be constant is if none of the vortex lines that make up the surface of the tube poke through the loop. So "A" must stay attached to the vortex tube, and since "A" is a material surface, the tube must be a material surface too.

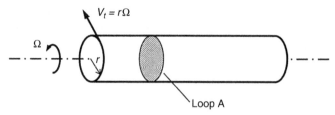

Figure 8.6 Kelvin's circulation theorem and Helmholtz's theorem on a vortex tube. *(Created by B. Cheung).*

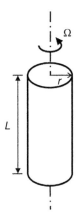

Figure 8.7 A vortex tube. *(Created by B. Cheung).*

Another important invariant is the angular momentum. For a circular vortex tube such as that shown in Fig. 8.7,

$$\text{angular momentum} = I\Omega \tag{8.14}$$

where the moment of inertia of the tube, $I = mr^2/2$, and the mass of the tube, $m = \rho\pi r^2 L$. We can rewrite the angular momentum as

$$\text{angular momentum} = mr^2\Omega/2 \tag{8.15}$$

or from Eq. 8.13

$$\text{angular momentum} = m\Gamma/(4\pi) \tag{8.16}$$

Since circulation is conserved and so is mass m for a material tube, the angular momentum is also a constant in the absence of external forces such as friction.

At this point, we can examine the changes that occur when a vortex tube undergoes distortion. Consider an axially stretched two-dimensional vortex tube as depicted in Fig. 8.8. In a barotropic flow, the vortex tube is a material surface, so the mass in the tube remains constant; that is

$$m = \rho\pi r^2 L = \text{constant} \tag{8.17}$$

This can be rewritten as

$$r^2 = \text{constant}/\rho\pi L \tag{8.18}$$

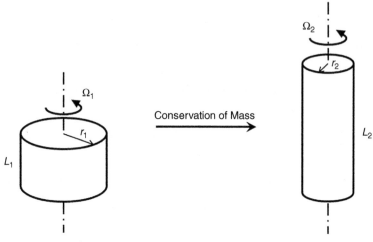

Figure 8.8 Invariant for vortex tube distortion. *(Created by B. Cheung).*

The most important invariant is probably the circulation, which can be expressed as

$$\Gamma = 2\pi r^2 \Omega = \text{constant} \tag{8.19}$$

Substituting for r^2 gives

$$\Gamma = 2\pi \left(\text{constant}/\pi\rho L\right)\Omega = \text{constant} \tag{8.20}$$

or

$$\Omega/\rho L = \text{constant} \tag{8.21}$$

Batchelor (1967) expressed this more elegantly in terms of vorticity, $\omega = 2\Omega$. The tangential velocity is related to the vorticity and the angular velocity via

$$V_t = r\Omega = r\omega/2 \tag{8.22}$$

Thus

$$\Omega = V_t/r = V_t/(\text{constant}/\pi\rho L)^{1/2} \tag{8.23}$$

Substitute this into $\Omega/\rho L = \text{constant}$, and we have

$$V_t/\rho_L (\text{constant}/\pi\rho L)^{1/2} = \text{constant} \tag{8.24}$$

Moving the constant on the left-hand side to the right, we have

$$V_t/(\rho L)^{1/2} = \text{constant} \tag{8.25}$$

or

$$V_t^2/\rho L = \text{constant} \tag{8.26}$$

This expression lucidly delineates that stretching (increasing) L leads to increasing tangential velocity V_t, and squashing (decreasing) L results in decreasing V_t. In other words, a vortex loses its vigor when being squashed and intensifies when being stretched.

8.4 EVOLUTION OF VORTICITY

A rigorously derived vorticity equation is due. We will follow Currie's (1974) detailed approach here. From the momentum equation for ρ = constant, we have

$$\frac{\partial \vec{U}}{\partial t} + \vec{U} \cdot \nabla \vec{U} = -\frac{1}{\rho} \nabla P + v \nabla^2 \vec{U} + \nabla \phi \tag{8.27}$$

We can expand the second term on the left-hand side via tensor identity

$$(\vec{a} \cdot \nabla)\vec{a} = \tfrac{1}{2}\nabla(\vec{a} \cdot \vec{a}) - \vec{a} \times (\nabla \times \vec{a}) \tag{8.28}$$

to get

$$\frac{\partial \vec{U}}{\partial t} + \nabla\left(\frac{1}{2}\vec{U} \cdot \vec{U}\right) - \vec{U} \times (\nabla \times \vec{U}) = -\frac{1}{\rho} \nabla P + v \nabla^2 \vec{U} + \nabla \phi \tag{8.29}$$

Taking curl ($\nabla \times$), we get, for v = constant

$$\frac{\partial \vec{\omega}}{\partial t} + \nabla \times \nabla\left(\frac{1}{2}\vec{U} \cdot \vec{U}\right) - \nabla \times (\vec{U} \times \vec{\omega}) = -\frac{\nabla \times \nabla P}{\rho} + \nabla \times (v \nabla^2 \vec{U}) + \nabla \times \nabla \phi \tag{8.30}$$

But $\nabla \times \nabla \varphi = 0$ for any scalar φ, and thus we are left with

$$\frac{\partial \vec{\omega}}{\partial t} - \nabla \times (\vec{U} \times \vec{\omega}) = \nabla \times (v \nabla^2 \vec{U}) \tag{8.31}$$

We note that the pressure term disappears, significantly easing the problem at hand. Therefore, the vorticity equation can be expressed as

$$\frac{\partial \vec{\omega}}{\partial t} - \nabla \times \left(\vec{U} \times \vec{\omega} \right) = v \nabla^2 \vec{\omega} \tag{8.32}$$

for ρ = constant and v = constant. Using vector identity, the second term can be expanded into

$$\nabla \times \left(\vec{U} \times \vec{\omega} \right) = \vec{U} \left(\nabla \cdot \vec{\omega} \right) - \vec{\omega} \left(\nabla \cdot \vec{U} \right) - \left(\vec{U} \cdot \nabla \right) \vec{\omega} + \left(\vec{\omega} \cdot \nabla \right) \vec{U} \tag{8.33}$$

We note that the divergence of the curl of a vector is zero, that is, $\nabla \cdot \vec{\omega} = 0$. In addition, the conservation of mass for the incompressible case leads to $\nabla \cdot \vec{U} = 0$. Therefore, the vorticity equation is

$$\frac{D\vec{\omega}}{Dt} = \frac{\partial \vec{\omega}}{\partial t} + \left(\vec{U} \cdot \nabla \right) \vec{\omega} = \left(\vec{\omega} \cdot \nabla \right) \vec{U} + v \nabla^2 \vec{\omega} \tag{8.34}$$

where the partial derivative term is the storage and the subsequent one is the convective term. This expression can be expanded into

$$\frac{D\vec{\omega}}{Dt} = \left(\vec{\omega} \cdot \nabla \right) \vec{U} + v \nabla^2 \vec{\omega} + \frac{\nabla \rho \times \nabla P}{\rho^2} - \vec{\omega} \left(\nabla \cdot \vec{U} \right) \tag{8.35}$$

plus additional terms if $v \neq$ constant.

For varying density flow where $\rho \neq$ constant, we see that

1.

$$\nabla \times \left(-\frac{1}{\rho} \nabla P \right) = -\frac{1}{\rho} \left(\nabla \times \nabla P \right) + \frac{1}{\rho^2} \nabla \rho \times \nabla P \tag{8.36}$$

where the first term on the right-hand side is zero since the curl of a gradient of a scalar is zero;

2.

$$-\vec{\omega} \left(\nabla \cdot \vec{U} \right) \neq 0 \tag{8.37}$$

where the term in the brackets is greater than zero for volume expansion; that is, volume expansion decreases the magnitude of vorticity;

3.

$$\nabla \times \left(v \nabla^2 \vec{U} \right) = \nabla^2 \vec{\omega} + \cdots \tag{8.38}$$

where the three dots on the right-hand side signify additional terms when μ or v is not a constant.

It is worth stressing that for constant density and viscosity flows, pressure does not appear explicitly. As such, the vorticity and velocity vectors may be obtained with no knowledge of pressure. In this case, the pressure acts through the center of gravity of each element, producing no vorticity.

Let us look at the tilting/stretching term. Knowing that

$$\vec{\omega} \cdot \nabla = \omega_x \frac{\partial}{\partial x} + \omega_y \frac{\partial}{\partial y} + \omega_z \frac{\partial}{\partial z} \tag{8.39}$$

we have

$$\vec{\omega} \cdot \nabla \vec{U} = \left(\omega_x \frac{\partial u}{\partial x} + \omega_y \frac{\partial u}{\partial y} + \omega_z \frac{\partial u}{\partial z} \right) \hat{i} + \left(\omega_x \frac{\partial v}{\partial x} + \omega_y \frac{\partial v}{\partial y} + \omega_z \frac{\partial v}{\partial z} \right) \hat{j}$$
$$+ \left(\omega_x \frac{\partial w}{\partial x} + \omega_y \frac{\partial w}{\partial y} + \omega_z \frac{\partial w}{\partial z} \right) \hat{k} \tag{8.40}$$

For example, Fig. 8.9a depicts the effect induced by tilting terms such as

$$\omega_y \frac{\partial u}{\partial y}, \omega_z \frac{\partial u}{\partial z}, \omega_x \frac{\partial v}{\partial x}, \omega_z \frac{\partial v}{\partial z}, \omega_x \frac{\partial w}{\partial x}, \omega_y \frac{\partial w}{\partial y} \tag{8.41}$$

Stretching of a vortex tube is illustrated in Fig. 8.9b. This axial stretching can be caused by

$$\omega_x \frac{\partial u}{\partial x}, \omega_y \frac{\partial v}{\partial y}, \omega_z \frac{\partial w}{\partial z} \tag{8.42}$$

The combined tilting and stretching can lead to the transition of the boundary layer as portrayed in Fig. 8.9c as a side view and in Fig. 8.9d as a plane view.

8.5 INTERPRETING TANGENTIAL VELOCITY AS TURBULENCE

We may model flow turbulence in terms of the three orthogonal vortical structures as shown in Fig. 8.10 (Wilson, 1989). The vortex tube lying with its axis along the x direction has a tangential velocity that will produce turbulence velocities v and w. Consider this vortex tube L_x, from $V_t^2/\rho L = \text{constant}$, we have

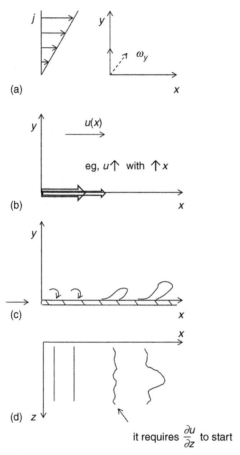

Figure 8.9 Tilting and stretching of vortex tubes: (a) $\partial u/\partial y > 0$, (b) $\partial u/\partial x > 0$, (c) side view of a boundary layer development, (d) plane view of a boundary layer development. *(Created by J. Smith).*

$$\frac{\overline{v_1^2}}{\overline{v_0^2}} = \frac{\overline{w_1^2}}{\overline{w_0^2}} = \frac{L_{x1}}{L_{x0}}\left(\frac{\rho_1}{\rho_0}\right) \tag{8.43}$$

This applies to all x-aligned vortex tubes in a barotopic flow with $\rho = \rho(P)$. It is clear that stretching this tube increases the fluctuating intensities in the y and z directions. Similarly, the stretching of the y-aligned vortex tube results in the intensification of u and w. And elongating the z-aligned vortex tube produces augmentation of u and v.

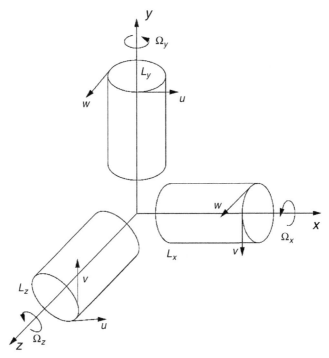

Figure 8.10 The three orthogonal vortex model. *(Created by N. Cao).*

8.5.1 Stretching of Vortex Tubes by Flow Acceleration

If the flow velocity U accelerates from U_0 to U_1, then the vortex tube L_x will be stretched as depicted in Fig. 8.11. In time Δt the length L will change by

$$\Delta L = \{[U + (dU/dx)L] - U\}\Delta t \qquad (8.44)$$

or

$$dL/dt = L\,dU/dx \qquad (8.45)$$

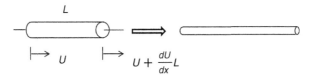

Figure 8.11 Stretching a streamwise vortex tube in the streamwise direction. *(Created by J. Smith).*

This can be rearranged into

$$dL/L = dU/(dx/dt) \qquad (8.46)$$

or

$$dL / L = dU / U \qquad (8.47)$$

Integrate this from L_0 to L_1, and we get

$$\ln(L_1 / L_0) = \ln(U_1 / U_0) \qquad (8.48)$$

or

$$L_1 / L_0 = U_1 / U_0 \qquad (8.49)$$

When a straight vortex tube is stretched axially, such as that shown in Fig. 8.11, the corresponding velocity components in the two orthogonal directions augment. This implies a corresponding enhancement in the turbulence intensity. If the stretching is in the x direction, then the increases in V and W signify intensification of v and w. The other outcome of the stretching is the decrease in vortex diameter, which represents the eddy size. Recall from Chapter 4 that both an increase in turbulence intensity and a decrease in eddy size serve to escalate the rate of dissipation. This tug-of-war can result in a higher or lower turbulence level, depending on the specific conditions involved. In general, the immediate outcome is elevated turbulence intensity, while the prolonged outcome is weakened turbulence.

8.5.2 Oblique Vortex Tubes Passing Through a Contraction

It is interesting to note that oblique vortex tubes can rotate, even in an irrotational flow. Let us examine two oblique vortex tubes passing through a two–dimensional contraction as depicted in Fig. 8.12. For incompressible flow, mass conservation requires that the volume of the fluid element remain unchanged. As such, the area of the two boxes in the figure is the same. If the flow is irrotational, we require only that the net rotation of the two vortex tubes A–A and B–B be equal and opposite. Specifically

$$\partial(\theta_A + \theta_B) / \partial x = 0 \qquad (8.50)$$

Nevertheless, the rotation of these oblique vortex tubes causes a redistribution of velocity components as some U becomes V. Thereupon, the distortion of oblique vortex tubes redistributes turbulence energy from $\overline{u^2}$ to $\overline{v^2}$.

Shear such as that in a boundary layer can produce similar effects as that just discussed. Figure 8.13 shows a vortex tube initially aligned with the

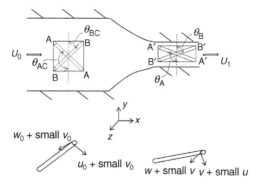

Distortion tilts u to become v

Figure 8.12 Oblique vortex tubes passing through a two-dimensional contraction. *(Created by J. Smith).*

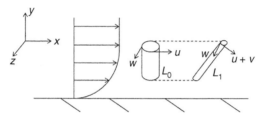

Figure 8.13 Reorientation of a vortex tube by shear flow. *(Created by J. Smith).*

y-axis, which is distorted by the boundary layer flow. As the mean flow is rotational, the effect in shear flow is relatively larger. We see that some of the energy associated with U and W is redistributed into V.

8.5.3 Compressing a Vortex Tube

Another way to distort a vortex tube is via compression or expansion. Figure 8.14 shows the compression of a vortex tube. We see that both the vortex tube length and the vortex core radius are reduced under compression. The shape and aspect ratio of the vortex tube itself, on the other hand, remain unchanged.

8.5.4 Vortex Tube Distortion by an Expanding Sphere

In a spark-ignition combustion chamber, the expanding flame can seriously alter the vortical structures in the chamber, and vice versa. Let us approximate the enlarging flame ball as a nonreacting expanding sphere in an open atmosphere where the pressure remains constant. In the ideal situation portrayed in Fig. 8.15, the two vortices parallel to the sphere surface are stretched while the one normal to the expanding sphere is squashed. The

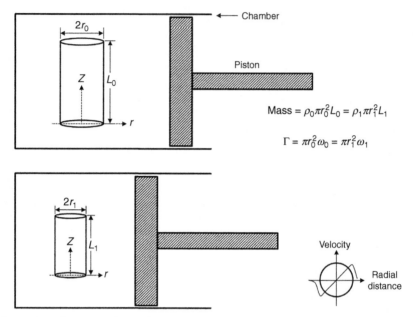

Figure 8.14 A vortex tube undergoing compression. *(Created by M. Ahmadi-Baloutaki based on Ting [1995]).*

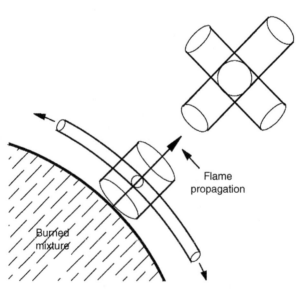

Figure 8.15 Three orthogonal vortex tubes in front of an expanding sphere in open atmosphere. *(Created by Ting [1995]).*

stretching of the two parallel vortex tubes leads to an increase in turbulence intensity, while the squashing of the normal one reduces the associated intensity perpendicular to the sphere surface. This geometric distortion is most intense immediately ahead of the growing sphere and diminishes farther away. Therefore, it can enhance the sphere front turbulence significantly, just as the sphere arrives. However, it has little effect on the overall turbulence decay rate in the flow field away from the sphere.

In reality, the combustion chamber is closed. As such, the expanding sphere in a closed vessel compresses the enclosed fluid, including the liquid far ahead of the sphere. This compression leads to increases in the turbulence intensity ahead of the sphere. The smaller, compressed vortical structures tend to also increase the turbulence decay rate. A theoretical estimate of spark-ignited, 70% stoichiometric methane-air (initially at atmospheric pressure and temperature) turbulent combustion inside a 2 L^3 combustion chamber with an equivalent radius of 76.6 mm (Ting, 1995) is portrayed in Fig. 8.16. For the purpose of this discussion, we simply treat the growing flame as a spherically expanding ball somewhat similar to the deployment of an air bag in a spherical enclosure with a radius of 76.6 mm. The (initial) turbulence was generated by passing a grid across the chamber; the resulting integral length was approximately 4 mm. The solid normal decay line illustrates normal turbulence decay with respect to time, without considering

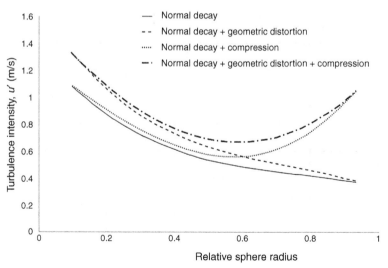

Figure 8.16 Estimated rapid distortion effects on an expanding sphere front turbulence, $\Lambda{\approx}4$ mm, chamber equivalent radius = 76.6 mm. *(Created by A. Goyal based on Ting [1995]).*

Figure 8.17 Fearful vorticity. *(Created by S.P. Mupparapu, edited by D. Ting).*

any effect of distortion. We note that (geometric) distortion for a freely expanding sphere, as discussed in the previous paragraph, only effectively applies to the region just ahead of the sphere. The dashed, normal decay plus
geometric distortion line accounts for the influence of sphere front distortion, in addition to normal decay. Note that sphere front distortion is most
pronounced when the expansion is greatest. This occurs when the sphere
is smallest where the chamber pressure is lowest and thus, least resistant to
the expansion. Considering the effect of compression, in addition to normal
decay, leads to the dotted normal decay plus compression line. Compression
is proportional to the chamber pressure; it is negligible initially and peaks at
the end as the pressure maximizes out. The dash-dotted, normal decay plus
geometric distortion plus compression line depicts the cumulative result en-
compassing normal decay, geometric or sphere front distortion, and compression. What has not been included is the presumably heightened decay rate.

It is thus clear that vortex dynamics plays a significant role in flow turbulence study. Like turbulence, vorticity at times must be feared and not
messed with; see Fig. 8.17.

Problems

Problem 8.1 Forced versus free vortex

A fluid in a circular container undergoes a rigid body rotation, that is,
$V_r = 0$ and $V_\theta = f(r)$, where r is the radial distance from the center. Deduce
the rotation, vorticity, and circulation. Is it possible to choose $f(r)$ so that the
flow is irrotational? How?

Problem 8.2 Rapid distortion of turbulence

A wind tunnel for generating isotropic turbulence has a symmetric contraction that decreases from a 1 m by 1 m into a 0.4 m by 0.4 m cross section. Calculate the change in the turbulence components u_{rms}, v_{rms}, and w_{rms} passing through the contraction. Also estimate the change in the turbulent kinetic energy and the relative turbulence intensities.

Problem 8.3 Vortex ring from an underwater balloon

The underwater balloon in Vasel-Be-Hagh et al. (2015) is assumed to be a perfect sphere where all its potential energy is converted into a smooth vortex ring propagating upward. Deduce the circulation and the upward propagation speed of the vortex ring at 3 m below the water level if the sphere is of 1 L and initially located at 7 m below the water level. You may first assume that the vortex ring expansion rate is negligible. What would some instabilities that corrugate the vortex ring do to the propagation speed?

REFERENCES

Batchelor, G.K., 1967. An Introduction to Fluid Dynamics. Cambridge, Cambridge.

Bernard, P.S., 2013. Vortex dynamics in transitional and turbulent boundary layers. AIAA J. 51 (8), 1828–1842.

Currie, I.G., 1974. Fundamental Mechanics of Fluids. McGraw-Hill, New York.

Helmholtz, H., 1858. Über integrale der hydrodynamischen gleichungen welche den wirbelbewegunden entsprechen, Journal für die Reine und Angewandte Mathematik, 55: 25–55. English translation: Tait, P.G., 1;1867. On integrals of the hydrodynamical equations which express vortex motion, Philosophical Magazine, 33: 485–512.

Kelvin, L., 1867a. The translatory velocity of a circular vortex ring (W. Thomson, Trans.). Philosophical Magazine 33, 511–512.

Kelvin, L., 1867b. On vortex atoms, Proceedings of the Royal Society of Edinburgh, VI: 94-105 (W. Thomson, Trans.). Reprinted in Philosophical Magazine, 34: 15–24.

Kelvin, L., 1868. On vortex motion, Transactions of the Royal Society of Edinburgh (W. Thomson, Trans.), 25: 217–260.

Kelvin, L., 1880. Vibrations of a columnar vortex (W. Thomson, Trans.). Philosophical Magazine Series 5, 10: 155–168.

Kundu, P.K., Cohen, I.M., Dowling, D.R., 2015. Fluid Mechanics, sixth ed. Academic Press, USA.

Prandtl, L., Tietjens, O.G., 1957. Fundamentals of Hydro- and Aeromechanics. Dover, New York.

Pritchard, P.J., Mitchell, J.W., 2015. Fox and McDonald's Introduction to Fluid Mechanics, ninth ed. Wiley, USA.

Pullin, D.I., Saffman, P.G., 1998. Vortex dynamics in turbulence. Ann. Rev. Fluid Mech. 30, 31–51.

Saffman, P.H., 1992. Vortex Dynamics. Cambridge University Press, USA.

Ting, D.S.-K., 1995. Modelling Turbulent Flame Growth in a Cubical Chamber, PhD Thesis, University of Alberta, Edmonton.

Tokaty, G.A., 1971. A History and Philosophy of Fluid Mechanics. Dover, New York.

Vasel-Be-Hagh, A.R., Carriveau, R., Ting, D. S.-K., 2015. A balloon bursting underwater. J. Fluid Mech. 769, 522–540.

White, F.M., 2006. Viscous Fluid Flow, third ed. McGraw-Hill, New York.
Wikipedia, 2015. https://en.wikipedia.org/wiki/Helmholtz%27s_theorems (accessed 03.07.2015.).
Wilcox, D.C., 2007. Basic Fluid Mechanics, third ed. DCW Industries, San Diego.
Wilson, D.J., 1989. Mec E 632 Turbulent Fluid Dynamics. University of Alberta, Edmonton.

PART 2

Examples of Engineering Problems Involving Flow Turbulence

CHAPTER 9

Sphere and Circular Cylinder in Cross Flow

The difference between something good and something great is attention to detail.

—Charles R. Swindoll

Contents

NOMENCLATURE

C_D	Drag coefficient, $= F_D/(\frac{1}{2}\rho U^2 D)$; where F_D = drag force
d, D	Diameter
f	Frequency
h	(Convective) heat transfer coefficient
k	Thermal conductivity
Nu	Nusselt number, convection/conduction, $= hD/k$
Re	Reynolds number, inertia force/viscous force, $= UD/\nu$
St	Strouhal number, $= fD/U$
t	A characteristic time period, a time scale
$TrBL$	Transition in the boundary layer
Tu	Turbulence intensity
U	Velocity

Greek Symbols

Λ	Large length scale; integral length
ν	Kinematic viscosity
ρ	Density

Basics of Engineering Turbulence
http://dx.doi.org/10.1016/B978-0-12-803970-0.00009-X

9.1 INTRODUCTION

A sphere or a circular cylinder in cross flow is one of the most basic scenarios in nature and in engineering applications. Furthermore, both spheres and circular cylinders furnish the simplest element of flow over a bluff body. No wonder these classic, more than a century old, fluid-structure beauties continue to attract so much attention; see Tan et al. (2005), Thompson et al. (2006), Almedeij (2008), Yeung (2008, 2009), Rodríguez et al. (2013), Fukada et al. (2014), and Cai and Sun (2015). Some of the recent endeavors remain focused on the flow and/or conventional aerodynamic control (Choi et al., 2008), while others venture into bio-mimicry such as those involving hydrophobic surfaces (You and Moin, 2007; Muralidhar et al., 2011). Among other objectives, reducing drag and vibration is a common practical goal behind many of these studies (Byon et al., 2010; Muddada and Patnaik, 2010; Gruncell et al., 2013). We will limit our discussion to the effect of free-stream turbulence on a smooth, conventional sphere and on a smooth, circular cylinder.

9.2 FLOW OVER A SMOOTH SPHERE

Laminar free stream over a smooth sphere is a classical example of flow over a bluff body. It is presumably the simplest three-dimensional bluff body because it is axisymmetric. Table 9.1 summarizes the flow characteristics for a smooth sphere in "laminar" flow (Tyagi et al., 2004, 2006), where "laminar" is assumed as having free-stream turbulence less than 0.5%. Strouhal number is defined as St $= fD/U$, where f is the vortex shedding frequency, D is the diameter of the sphere, and U is the free-stream velocity. The key information utilized in creating this summary table came from Achenbach (1972) and Taneda (1978). The critical Reynolds number is defined as the Reynolds number at which the drag coefficient C_D undergoes a sudden drop, after remaining roughly constant over a wide range of Re, as shown in Fig. 9.1. This abrupt reduction in drag is associated with a leeward shift of the separation location (circle). The main features associated with the flow around a sphere are sketched in Fig. 9.2.

It is relatively well-accepted that an increase in flow turbulence advances the laminar-to-turbulent boundary layer transition to a lower Reynolds number compared to its "smooth flow" counterpart. This transition reduces the adverse pressure gradient around the sphere, delaying the separation point farther downstream. As a result, the pressure drag is progressively lowered with increasing free-stream turbulence. The most noticeable outcome

Table 9.1 Flow characteristics of a smooth sphere in laminar free stream.

Characteristic	$Re < 20$	$20 < Re < 400$	$400 < Re < 10^3$	$10^3 < Re < 3 \times 10^3$	$3 \times 10^3 < Re < 6 \times 10^3$	$6 \times 10^3 < Re < 3 \times 10^5$	$3 \times 10^5 < Re < 5 \times 10^6$
Characteristic	Potential flow	Vortex ring	Vortex loops	–	Lower critical Re	–	Higher critical Re
Boundary layer	Laminar	Laminar	Laminar	Laminar	Laminar	Laminar	Turbulent
Wake	–	Negligible periodic fluctuations	Strong periodic fluctuations	Strong periodic fluctuations	Strong periodic fluctuations	Strong periodic fluctuations	Stop fluctuating periodically
St	–	~0.2	~0.2	~0.2	~0.2, ~2.0	~0.2, ~2.0	–
	No separation	A stationary vortex ring on the leeward side	The vortex ring stretched into vortex loops	Vortex loops diffuse into the wake	St–Re discontinuity @ $Re\sim6 \times 10^3$	Separation point shifts leeward	C_D–Re discontinuity @ $Re\sim3 \times 10^5$

Source: Based on Achenbach (1972) and Taneda (1978).

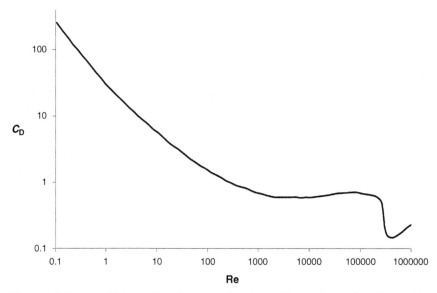

Figure 9.1 Drag coefficient of a sphere as a function of Reynolds number. *(Created by A. Goyal).*

of this is the advancement of the critical Reynolds number associated with this drag crisis. Among many others, Torobin and Gauvin's (1961) experimental results clearly portray the progressively lowered critical Reynolds number with increasing turbulence intensity. This effect also appears to be more obvious at higher or moderately high Re.

Furthermore, the wind tunnel experiments of Moradian et al. (2009, 2011) appear to concur with studies such as that of Savkar et al. (1980) that conclude turbulence with integral length scale of size equal to or somewhat less than the diameter of the bluff body is more effective in advancing the drag crisis. Figure 9.3 summarizes the effects of Reynolds number, turbulence intensity and integral length scale on the drag coefficient of a smooth sphere, as deduced by Moradian et al. (2009, 2011). It is worth mentioning that the corresponding standard C_D value over this range of Re is around 0.5. The results seem to show that a higher level of turbulence always leads to a lower C_D, notwithstanding the fact that the range of studied condition is limited. More interestingly, for the same relative turbulence intensity at a particular Re, integral length scale Λ of approximately the diameter (D) of the sphere leads to the lowest C_D.

As mentioned earlier, one key role of free-stream turbulence is in advancing the laminar–to–turbulent boundary layer transition. The underlying

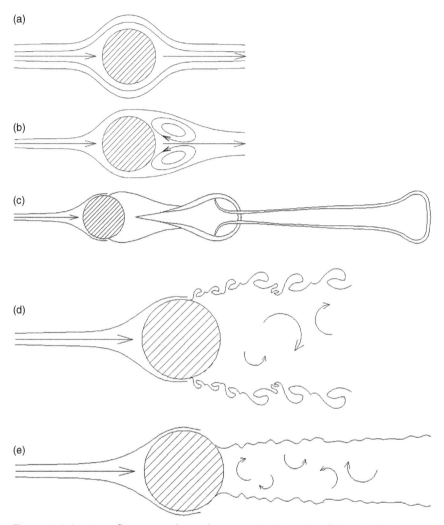

Figure 9.2 Laminar flow around a sphere at: (a) Re < 20; (b) 20 < Re < 400; (c) 400 < Re < 10³; (d) 10³ < Re < 3 × 10³; (e) Re > 3 × 10³. *(Created by H. Tyagi).*

physics associated with the integral length scale effect, however, is less obvious. There are at least two important dimensions concerning flow over a sphere: the thickness of the boundary layer, especially that just upstream of the separation point; and the size of the wake, which may be approximated by the diameter of the sphere. The sphere results presented in Moradian et al. (2009, 2011) seem to suggest that, over the range of studied conditions, turbulent flow with eddy sizes which fall in around the boundary-layer thickness and wake size is most effective in advancing the drag crisis.

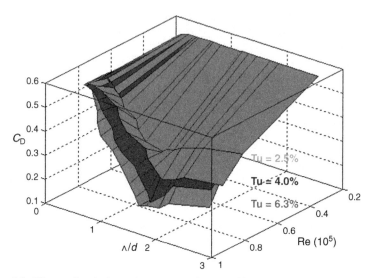

Figure 9.3 Effects of turbulence intensity and integral length scale on the C_D-Re curve of a smooth sphere. *(Created by N. Moradian).*

9.3 SMOOTH FLOW ACROSS A CIRCULAR CYLINDER

For a smooth circular cylinder in a smooth (non-turbulent) cross flow, the general flow regimes are relatively well-defined (Blevins, 1990; Zdravkovich, 1997). However, the details and the subdivisions within these extensively studied flow regimes are still subject to debates and are still being scrutinized by many researchers today. This is particularly true at Reynolds numbers in excess of a few hundred thousand, where our understanding of the complex fluid mechanics and steady and unsteady loading on the cylinder is far from complete (Zan, 2008). Figure 9.4 portrays the typical C_D versus Re plot. The general flow regimes, especially for Reynolds numbers less than a million, can be more or less described in the following manner:

Regime 1: Creeping Flow (Re ≤ 5).

For Reynolds numbers of less than approximately five, the streamlines firmly attach around the cylinder circumference with no visible wake on the leeward side. Thus, this flow regime is referred to as the regime of unseparated flow.

Regime 2: Steady, Closed Near-Wake (5 ≤ Re ≤ 45).

When the Reynolds number is increased to larger than roughly five, a pair of standing vortices is formed in the near-wake. This fixed pair of vortices is commonly known as Föppl vortices.

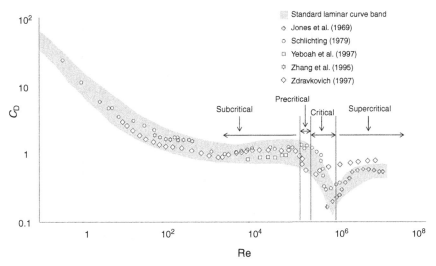

Figure 9.4 Drag coefficient of a circular cylinder as a function of Reynolds number. *(Created by A. Goyal based on data from Younis and Ting [2012]).*

Regime 3: Laminar Vortex Street ($45 \leq Re \leq 190$).

As the near-wake elongates and becomes unstable with increasing Reynolds number, a sinusoidal oscillation of the shear layers is initiated at the confluence point. The amplitude of this trail oscillation increases with increasing Re, until the well-known von Kármán (1911, 1912) vortex street is formed. According to Zdravkovich (1997), Bénard should also be credited for this discovery; that is, this vortex street should be called von Kármán–Bénard vortex street.

Regime 4: Transition in Shear Layers ($190 \leq Re \leq 3 \times 10^5$).

The laminar vortices in the wake become irregular and distorted when the Reynolds number is increased beyond a value of approximately 190. These nonlaminar vortical structures are formed by the rolling up of transition waves. When Re reaches a large enough value, there is a sudden burst into turbulence in the free shear layer near the cylinder. Consequently, turbulent eddies are formed near the rear of the cylinder. The pre-critical regime, characterized by the intrinsically three-dimensional near-wake, comes into existence at Re of around 3×10^5.

Regime 5: Critical Regime ($3 \times 10^5 \leq Re \leq 3 \times 10^6$).

This regime may be further divided into sub-regimes, from sub-critical to super-critical. A single separation bubble is formed on one side of the cylinder; this bias is portrayed as a preferential lift. With a further increase in Re, the second separation bubble emerges, and the overall flow returns to symmetry. This regime is characterized by a serious drop in the drag coefficient. This drastic drag reduction is attributed to the transition of the laminar boundary layer into a turbulent one. Associated with the turbulent boundary layer is the narrowing of the wake and the cessation of regular vortex shedding.

Regime 6: Post-Critical Regime (Re $\geq$ 3 $\times$ 10^6).

This regime is also called the trans-critical regime. It was a surprise discovery by Roshko (1961), who found the reestablishment of turbulent vortex street when Re exceeds one million.

Note that for flows with Reynolds numbers larger than about 10^5 (Regimes 4, 5, and 6 covered above), the flow regimes can be categorized according to the physical state of the boundary layer. The transition in the boundary layer (TrBL) can start at Re of 1 $\times$ 10^5, with an upper bound of roughly 5 $\times$ 10^6. Following this description approach, we can further sub-divide the TrBL regime into TrBL0, where the drag is significantly decreasing with separation points moving leeward: TrBL1, where a laminar bubble is formed on one side of the cylinder; TrBL2 with two laminar bubbles behind the cylinder and hence, return of flow symmetry; TrBL3, where the bubbles are disrupted in the span-wise direction; and TrBL4, where the bubbles are eliminated.

9.4 A CIRCULAR CYLINDER IN TURBULENT CROSS FLOW

The general notion from over a century of intensive research is that the shape of the C_D versus Re curve is not significantly altered in the presence of turbulence. Free-stream turbulence affects the aerodynamics mostly in shifting the flow regimes downward, that is, increasing the effective Re (Fage and Warsap, 1929; Kiya et al., 1982; Mulcahy, 1984; Sanitjai and Goldstein, 2001; Ai et al., 2013). Ohya (2004) deduced the drag coefficient of a circular cylinder in an extremely high-turbulence atmospheric flow (typhoon) and found the corresponding C_D values roughly equal to those measured in smooth wind tunnel flows. Zan (2008) used two grids to generate 5% and 33 mm integral length, and 13% and 74 mm integral length wind in a pressurized wind tunnel. With circular cylinders of 38, 75, and 150 mm diameters, Zan managed to vary the Reynolds number from 1 $\times$ 10^5 to 2 $\times$ 10^6 in the presence of the grid. Zan found that for the 5% turbulence case, the Strouhal number (St) first increases from about 2.7 to 3.3, followed by a region without coherent shedding, and re-emergence of shedding with St falling to the typical 0.2 value at Re above 2 $\times$ 10^6, with increasing Re. Thus, Zan concluded that free-stream turbulence promoted the return of strong coherent shedding at Reynolds numbers significantly lower than that for smooth flow.

It is most interesting to note that Ohya (2004) suggests that when the free-stream turbulence has (length) scales orders of magnitude larger than those associated with the cylindrical structure under consideration, the free-stream turbulence loses its relevance. Zan (2008) unfortunately did not examine the potential role played by the integral length. This is possibly, at least in part, because he was limited by only two values of integral length; nonetheless, with the three different cylinder sizes, there were six different integral length-cylinder diameter ratios to play with.

Other than the well-accepted augmentation of effective Re with increasing turbulence intensity, there is a lack of a consensus in the open literature concerning other details. These details include: (1) the possibility of very different effects of free-stream turbulence in different (smooth) flow regimes; (2) the dissimilar physics when dealing with a low, moderate, and high level of turbulent flow; and (3) the unique role of turbulence length scale. Concerning the role of turbulence length scale, many studies have found that its effect is inconsequential when compared to the turbulence intensity effect. There are, nevertheless, quite a few treasured exceptions amidst the copious publications.

Younis and Ting's (2012) circular cylinder results, as plotted in Figs 9.5 and 9.6, very much concur with Moradian et al. (2009, 2011) sphere in turbulent flow findings. Note that the corresponding value of C_D for the

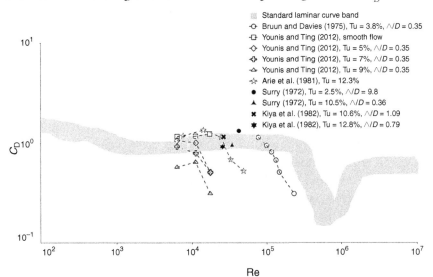

Figure 9.5 Drag coefficient of a circular cylinder with respect to Reynolds number under the influence of free-stream turbulence. *(Created by A. Goyal based on data from Younis and Ting [2012]).*

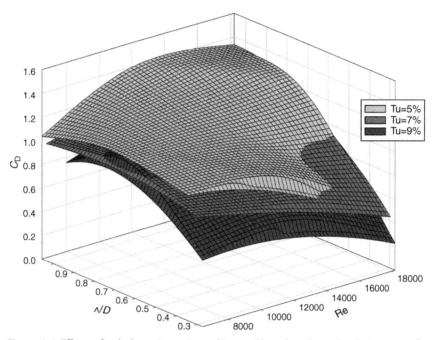

Figure 9.6 Effects of turbulence intensity and integral length scale on the C_D-Re curve of a smooth circular cylinder. *(Created by A. Ahmed based on data from Younis and Ting [2012]).*

standard nonturbulent flow over a smooth circular cylinder is around 1.2. Over the range of conditions considered, the most effective drag-reducing turbulence is that which has the highest intensity and smallest eddy size at the highest Reynolds number. The optimum drag reduction relative integral length for the circular cylinder case seems to be smaller than that for the sphere; that is, the smallest Λ/d of about 0.35 tested in the cylinder case as compared to Λ/d of around unity in the sphere study. Another interesting observation not found in Moradian et al. (2009, 2011), but showed up quite consistently in Younis and Ting (2012) is the relatively high "smooth flow" C_D value at moderately low relative turbulence intensity of around 5%, and with Λ/d of around unity. This hump, which occurs just before the turbulence-advanced drop in C_D has also been observed by researchers such as Savkar et al. (1980).

9.5 TURBULENT FLOW OVER A HEATED CIRCULAR CYLINDER

Turbulent forced convection from a circular cylinder has been extensively researched. Nevertheless, the general consensus concerning the effect of turbulence is still only limited to the qualitative trend of increasing Nusselt number (Nu = hD/k; where h is the heat transfer coefficient, D is the diameter of the cylinder, and k is the thermal conductivity of the fluid) with increasing Re. There is an appreciable measure of scatters (in excess of ±50%) in the data on the amount of heat transfer enhancement induced by the free-stream turbulence. We believe that a significant portion of this scatter is due to the very different turbulence generated in various studies, and that the level of turbulence in addition to Re alone is inadequate in specifying the dissimilar turbulence encountered from one study to another. Even for the simplest form of turbulence, the quasi-isotropic turbulence generated by a well-designed orificed grid (Liu and Ting, 2007), both the turbulence intensity and the integral length scale are required to formulate a minimum description for the turbulent flow. Systematic studies on the effect of eddy size on convection heat transfer are very scarce in the literature; on the contrary, there are many published papers which do not have a clear basic understanding of turbulence claiming that the role of integral length is non-consequential. We wish to single out van der Hegge Zijnen (1958) and Žukauskas et al. (1993), for they are some of the very few researchers who have systematically and successfully scrutinized the subtle role of turbulent length scale. Both studies agree that there is some sort of optimum Λ/d at which Nu peaks, at a given Re and turbulence intensity. This optimal Λ/d value deduced by van der Hegge Zijnen (1958) over the range of flow conditions considered is approximately 1.5, whereas that obtained by Žukauskas et al. (1993) over the conditions they explored is significantly smaller at around 0.1. Figure 9.7 compares the results obtained by Sak et al. (2007) with Žukauskas et al. (1993) study; keep in mind that the conditions such as Re and turbulence intensity differ among these studies. We see that while the values of relative Nusselt number obtained by Sak et al. (Re = 2.8×10^4 and 6.7% turbulence) are lower than that of Žukauskas et al., their qualitative trends are quite agreeable.

In short, forced convection heat transfer around a cylinder is very much dictated by the flow around the bluff body. We expect turbulence of eddies on the order of the boundary layer and of the wake to have the largest impact on the rate of heat transfer. We are not aware of any study, which finds an attenuation of heat transfer rate with the introduction of turbulence. Therefore, we believe that turbulence of integral length on the order of boundary-layer thickness is likely most effective in augmenting the heat transfer. Eddies of the same size as the wake are also likely of significance and may alter the wake and potentially the separation point, as well.

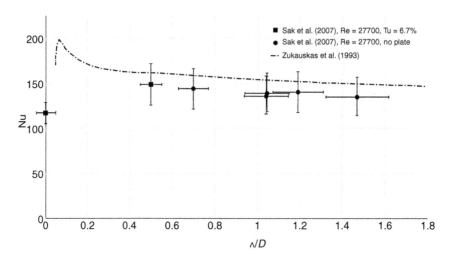

Figure 9.7 The subtle role of turbulent integral length scale on the Nu-Re relationship for a smooth circular cylinder. *(Created by A. Goyal based on data from Sak et al. [2007]).*

9.6 SOME COMMENTS ON FLOW OVER A BLUFF BODY

To conclude, it is clear that there remains much work to be done in scrutinizing the various roles of the underlying turbulence parameters in bluff body aerodynamics and in convection heat transfer from a simple bluff body such as a smooth circular cylinder. The minimum requirement for conducting this type of research is some quasi-independent manipulation of the key turbulence parameters, in addition to controlling the Reynolds number independently.

Problems

Problem 9.1. Flow across a row of square plates
Four 7 cm $\times$ 7 cm plates are placed parallel to a 1 m/s water (at 20 °C) flow. In one configuration, the four plates are placed one after another, forming a 7 cm $\times$ 28 cm rectangular plate with its longer dimension (28 cm) parallel to the flow. In another configuration, there is a spacing of 3.5 cm between consecutive plates. Calculate the total drag of both configurations by assuming (1) laminar boundary layer flow, (2) turbulent boundary layer flow.

Problem 9.2. Dragging a sphere in water
A 1 m sphere of specific gravity of 0.5 is fully submerged in 20°C water. It is dragged by an underwater vehicle by a thin 2 m long wire. The wire makes a 45° angle with respect to the horizontal plane. What is the speed of the sphere?

Problem 9.3. Dimple versus flow agitation

You have a choice to create dimples on a sphere or to agitate the free stream with eddies of equivalent size. Which way is more effective in advancing the drag crisis? Backup your answer as rigorously as possible.

REFERENCES

Achenbach, E., 1972. Experiments on the flow past spheres at very high Reynolds numbers. J. Fluid Mech. 54 (3), 565–575.

Ai, Y., Feng, D., Ye, H., Li, L., 2013. Unsteady numerical simulation of flow around 2-D circular cylinder for high Reynolds numbers. J. Mar. Sci. Appl. 12, 180–184.

Almedeij, J., 2008. Drag coefficient of flow around a sphere: matching asymptotically the wide trend. Powder Technol. 186, 218–233.

Arie, M., Kiya, M., Suzuki, Y., Hagino, M., Takahashi, K., 1981. Characteristics of circular cylinders in turbulent flows. B. Jpn. Soc. Mech. Eng. 24 (190), 640–647.

Blevins, R.D., 1990. Flow-Induced Vibrations, second ed. Van Nostrand Reinhold, New York.

Bruun, H.H., Davies, P.O.A.L., 1975. An experimental investigation of the unsteady pressure forces on a circular cylinder in a turbulent cross flow. J. Sound Vib. 40 (4), 535–559.

Byon, C., Nam, Y., Kim, S.J., Ju, Y.S., 2010. Drag reduction in Stokes flows over spheres with nanostructured superhydrophilic surfaces. J. Appl. Phys. 107 (6), 1–3, 066102.

Cai, C., Sun, Q., 2015. Near continuum flows over a sphere. Comput. Fluids 111, 62–68.

Choi, H., Jeon, W.-P., Kim, J., 2008. Control of flow over a bluff body. Ann. Rev. Fluid Mech. 40, 113–139.

Fage, A., Warsap, J.H., 1929. The effects of turbulence and surface roughness on drag of a circular cylinder, British Aerodynamics Research Council, Report Memorandum 1283.

Fukada, T., Takeuchi, S., Kajishima, T., 2014. Effects of curvature and vorticity in rotating flows on hydrodynamic forces acting on a sphere. Int. J. Multiphase Flow 58, 292–300.

Gruncell, B.R.K., Sandham, N.D., McHale, G., 2013. Simulations of laminar flow past a superhydrophobic sphere with a drag reduction and separation delay. Phys. Fluids 25, 1–16, 043601.

Jones, G., Cinotta, J., Walker, R., 1969. Aerodynamic forces on a stationary and oscillating circular at high Reynolds number, NACA TR R-300.

Kiya, M., Suzuki, Y., Arie, M., Hagino, M., 1982. A contribution to the free-stream turbulence effect on the flow past a circular cylinder. J. Fluid Mech. 115, 151–164.

Liu, R., Ting, D.S-K., 2007. Turbulent flow downstream of a perforated plate: sharp-edged orifice versus finite-thickness holes. J. Fluids Eng. 129, 1164–1171.

Moradian, N., Ting, D.S-K., Cheng, S., 2009. The effects of freestream turbulence on the drag coefficient of a sphere. Exp. Therm. Fluid Sci. 33, 460–471.

Moradian, N., Ting, D.S-K., Cheng, S., 2011. Advancing drag crisis of a sphere via the manipulation of integral length scale. Wind Struct. 14 (1), 35–53.

Muddada, S., Patnaik, B.S.V., 2010. An active flow control strategy for the suppression of vortex structures behind a circular cylinder. Eur. J. Mech. B-Fluid 29, 93–104.

Mulcahy, T.M., 1984. Fluid forces on a rigid cylinder in turbulent cross flow, ASME Symposium on Flow-Induced Vibrations, Winter Annual Meeting, New Orleans, pp. 15–28.

Muralidhar, P., Ferrer, N., Daniello, R., Rothstein, J.P., 2011. Influence of slip on the flow past superhydrophobic circular cylinders. J. Fluid Mech. 680, 459–476.

Ohya, Y., 2004. Drag of circular cylinders in the atmospheric turbulence. Fluid Dyn. Res. 34 (2), 135–144.

Rodríguez, I., Lehmkuhl, O., Borrell, R., Oliva, A., 2013. Flow dynamics in the turbulent wake of a sphere at sub-critical Reynolds numbers. Comput. Fluids 80, 233–243.

Roshko, A., 1961. Experiments on the flow past a circular cylinder at very high Reynolds number. J. Fluid Mech. 10, 345–356.

Sak, C., Liu, R., Ting, D.S-K., Rankin, G.W., 2007. The role of turbulence length scale and turbulence intensity on forced convection from a heated horizontal circular cylinder. Exp. Therm. Fluid Sci. 31, 279–289.

Sanitjai, S., Goldstein, R.J., 2001. Effects of free stream turbulence on local mass transfer from a circular cylinder. Int. J. Heat Fluid Flow 44, 2863–2875.

Savkar, S.D., So, R.M.C., Litzinger, T.A., 1980. Fluctuating lift and drag forces induced on large span bluff bodies in a turbulent cross flow, ASME Heat Transfer Division, HTD 9, 19–26.

Schlichting, H., 1979. Boundary Layer Theory, seventh ed. McGraw-Hill, USA.

Surry, D., 1972. Some effects of intense turbulence on the aerodynamics of a circular cylinder at subcritical Reynolds number. J. Fluid Mech. 52, 543–556.

Tan, B.T., Thompson, M.C., Hourigan, K., 2005. Evaluating fluid forces on bluff bodies using partial velocity data. J. Fluid Struct. 20, 5–24.

Taneda, S., 1978. Visual observations of the flow past a sphere at Reynolds numbers between 10^4 and 10^6. J. Fluid Mech. 85 (1), 187–192.

Thompson, M.C., Hourigan, K., Ryan, K., Sheard, G.J., 2006. Wake transition of two-dimensional cylinders and axisymmetric bluff bodies. J. Fluid Struct. 22, 793–806.

Torobin, L.B., Gauvin, W.H., 1961. The drag coefficients of single spheres moving in a steady and accelerated motion in a turbulent fluid. Am. Inst. Chem. Eng. J. 7 (4), 615–619.

Tyagi, H., Liu, R., Ting, D.S-K., Johnston, C.R., 2004. Experimental investigation of perforated plate turbulent flow past a solid sphere, 2004 ASME International Mechanical Engineering Congress, Paper IMECE2004-60340.

Tyagi, H., Liu, R., Ting, D.S-K., Johnston, C.R., 2006. Measurement of wake properties of a sphere in freestream turbulence. Exp. Therm. Fluid Sci. 30, 587–604.

van der Hegge Zijnen, B.G., 1958. Heat transfer from horizontal cylinders to a turbulent airflow. Appl. Sci. Res. 7A, 205–223.

von Kármán, T., 1911. On the mechanism of drag generation on the body moving in fluid (in German), Nachrichten Gesellschaft Wissenschaften, Göttingen, Part 1: 509–517, Part 2: 547–556.

Yeboah, E.N., Rahai, H.R., LaRue, J.C., 1997. The effects of external turbulence on mean pressure distribution, drag coefficient, and wake characteristics of smooth cylinders, ASME Fluids Engineering Division Summer Meeting, FEDSM'97-3167.

Yeung, W.W.H., 2008. Self-similarity of confined flow past a bluff body. J. Wind Eng. Ind. Aerod. 96, 369–388.

Yeung, W.W.H., 2009. On pressure invariance, wake width and drag prediction of a bluff body in confined flow. J. Fluid Mech. 666, 321–344.

You, D., Moin, P., 2007. Effects of hydrophobic surfaces on the drag and lift of a circular cylinder. Phys. Fluids 19, 1–4, 081701.

Younis, N., Ting, D.S-K., 2012. The subtle effect of integral scale on the drag of a circular cylinder in turbulent cross flow. Wind Struct. 15 (6), 463–480.

Zan, S.J., 2008. Experiments on circular cylinders in crossflow at Reynolds numbers up to 7 million. J. Wind Eng. Ind. Aerod. 96, 880–886.

Zdravkovich, M.M., 1997. Flow Around Circular Cylinders Vol 1: Fundamentals. Oxford University Press, Oxford.

Zhang, H.-Q., Fey, U., Noack, B.R., Konig, M., Eckelmann, H., 1995. On the transition of the cylinder wake. Phys. Fluids 7 (4), 779–794.

Žukauskas, A., Vaitiekūnas, P., Žiugžda, J., 1993. Analysis of influence of free stream turbulence intensity and integral length scale on skin friction and heat transfer of a circular cylinder, experimental heat transfer. Fluid Mech. Therm., 591–596.

CHAPTER 10

Premixed Turbulent Flame Propagation

I consider nature a vast chemical laboratory in which all kinds of composition and decompositions are formed.

–Antoine Lavoisier

Contents

NOMENCLATURE

A	Area
Da	Damköhler number, Da = characteristic turbulence time scale/characteristic chemical time scale
h	Height
K_L	Karlovitz stretch, K_L = local flame residence time/laminar flame stretching time
KE	Kinetic energy
L	Length
Le	Lewis number, Le = rate of energy (heat) transport/rate of mass transport
m_l'	Mass burning rate of a laminar flame
m_t'	Mass burning rate of a turbulent flame
Ma	Markstein number, sensitivity of flame speed to stretch

Basics of Engineering Turbulence
http://dx.doi.org/10.1016/B978-0-12-803970-0.00010-6

P Pressure
pdf   Probability density function
r Radius; subscripts "i" and "o" signify inner and outer, respectively
r_m Mean flame radius
r_{um} Mean flame radius of a sphere enclosing the burned volume
R Radius
R_t Turbulent flame propagation speed
Re   Reynolds number
rms   Root mean square
rpm   Revolutions per minute
S_l Laminar flame speed, laminar burning velocity
S_f Flame propagation rate
S_t Turbulent flame speed
SI   Spark ignition
STP   Standard temperature and pressure
t Time
T Temperature
T_b Burned temperature
T_u Unburned temperature
u' Turbulence intensity
V Velocity
V_b Velocity of the burned gas
V_u Velocity of the unburned gas
x Thickness, distance in the x direction

Greek and other symbols

Δx_b Thickness of the burned mixture element
Δx_u Thickness of the unburned mixture element
δ_l Laminar flame front thickness
η Kolmogorov microscale
Λ Large integral length
λ Taylor microscale
ρ Density; ρ_b is the burned mixture density, ρ_u is the unburned mixture density
τ_{chem} Chemical time scale
τ_{flow} Flow time scale
Ω Angular velocity
ω Vorticity
$\forall$ Volume

10.1 INTRODUCTION

At this point we are quite comfortable with the general notions of laminar and turbulent flows. For laminar flows, the adjacent layers of fluid slide past one another in a smooth, orderly manner. The mixing is due to molecular diffusion. In turbulent flows, eddies move randomly in all directions, crossing adjacent fluid layers and significantly enhancing mixing.

The flows in most practical combustion devices are turbulent. Furthermore, turbulence is often designedly created and/or enhanced in order to augment the mass-burning rate. This increase in the chemical energy release rates in turn boost the power output. For example, as the speed (rpm) of an internal combustion engine is increased, the turbulence intensity increases, and thus the mass burning rate. For this reason, parameters such as spark timing do not have to be drastically altered as the engine speed changes.

What is turbulent combustion? It is somewhat difficult to provide a definite, rigorous answer to this. Nonetheless, in a general sense, we may view turbulent combustion simply as combustion characterized by turbulent flow. By the same token, laminar combustion is combustion, which takes place in a laminar environment. We are going to focus only on premixed flames in which the fuel and the oxidizer are well-mixed prior to ignition. Before going further, let us brief ourselves with some of the basic terminologies involved. Keep in mind that the purpose of this chapter is to illustrate flow turbulence in applications.

10.1.1 Premixed Laminar Flame

For a premixed laminar flame, it is possible to define a flame velocity that, within reasonable limits, is independent of the experimental apparatus. In other words, the laminar flame speed or laminar burning velocity depends only on the fuel, oxidizer, and transport properties such as thermal conductivity, viscosity, and molecular diffusivity. We will see that to define a turbulent flame velocity in such a rigorous manner is not possible.

The laminar flame speed or the laminar burning velocity

$$S_l = dx/dt \qquad (10.1)$$

is well-defined, at least theoretically, as portrayed in Fig. 10.1. This laminar flame speed is simply the speed of the unburned mixture entering the flame front. We note that multiplying this laminar flame speed with the surface area of the flame (combustion wave) gives the volumetric burning rate.

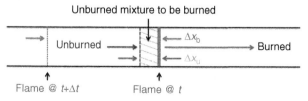

Figure 10.1 A planar, premixed, laminar flame. *(Created by D. Ting).*

With the accepted reference at standard temperature and pressure (STP), this can easily be converted into mass burning rate, the primary parameter of practical concern. In other words, laminar flame speed is the parameter that depicts the fundamental burning rate that is defined by the chemistry and thermodynamic state.

Returning to Fig. 10.1, the thickened element of unburned mixture of thickness Δx_u is consumed over an infinitely small time step dt; hence, $S_1 = \Delta x_u / dt$. We note that the organized one-dimensional laminar flow does not alter the chemistry. Accordingly, the laminar flame speed depends uniquely on the thermal and chemical properties of the mixture and the thermodynamic state defined by T and P. After the unburned mixture (reactants) in element Δx_u combusts, the resulting products are at a much higher temperature. Consequently, the element expands from Δx_u into Δx_b, where the thickness ratio $\Delta x_b / \Delta x_u$ is defined by the corresponding density ratio ρ_u / ρ_b. In the ideal situation with the right end of the tube closed and that on the left open in Fig. 10.1, the flame can only progress to the left. Under this condition, the laminar flame speed can be deduced from the propagation speed, which is the rate at which the flame travels. Specifically

$$S_1 = (\rho_b / \rho_u)\, dx_b / dt \qquad (10.2)$$

Here, dx_b / dt is the flame propagation speed.

10.1.2 Premixed Turbulent Flame

Under the influence of flow turbulence, the flame front may be distorted, as depicted in Fig. 10.2. We may define an equivalent flame speed, which is somewhat analogous to the laminar flame speed defined above. The right vertical dashed line in Fig. 10.2 symbolizes the mean location of the wrinkled turbulent combustion wave, which is portrayed by the wrinkled line on the right. Note that the wrinkled flame front is smooth on the unburned side, while it is rather pointed on the burned side. This is due to the comparatively faster volume or mass consumption in the valleys next to the

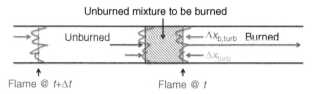

Figure 10.2 A one-dimensional, premixed, turbulent flame. *(Created by D. Ting).*

reacting flame front. The burning from the two sides of the valleys result in much steeper valleys with pointed ends toward the burned mixture. Over an infinitely small time dt, the volume of unburned mixture enclosed by the two wrinkled lines on the right is to be consumed. The vertical dashed lines enclose a volume (the hatched cross-sectional area for the two–dimensional case shown), which is equal to that bounded by the two wrinkled lines on the right. In other words, the hatched area defined by $\Delta x_{\text{turbulent}}$ multiplied by the depth of the channel encompasses the equivalent unburned volume of the mixture to be burned in an infinitesimal time step dt. As such, $\Delta x_{\text{turbulent}}/dt$ is the turbulent flame speed, which is also referred to as turbulent burning velocity. In other words, the turbulent flame speed

$$S_t = dx/dt \tag{10.3}$$

where the flame front can be smooth, wrinkled, distorted, or undefined. Note that the smooth case is the basic well-defined laminar flame, while the flame front or combustion wave can become undefined at a very high level of turbulence and/or very low laminar flame speed where there may be pockets of unburned and partially burned mixture.

It is useful at this point to also define the equivalent turbulent flame speed for a spherical flame, which is often encountered in an internal combustion engine. Let us start with the unconfined case where the pressure remains constant as the combustion takes place; that is, the flame is free to propagate and/or expand. The innermost corrugated line in Fig. 10.3 represents the turbulent flame front at time zero. The equivalent spherical volume containing the same burned mixture is represented by the innermost circle of radius r_m. Over an infinitely small time step dt, the unburned

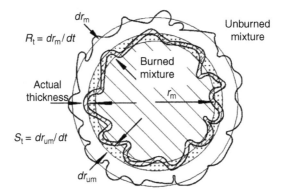

Figure 10.3 A premixed turbulent flame ball in open atmosphere. *(Created by D. Ting).*

mixture enclosed by the most inner and the adjacent corrugated lines is to be burned. This volume is equal to the shell enclosed by the innermost circle (sphere) of r_m and the next one, that is, the hatched element of thickness dr_{um}. The turbulent flame speed

$$S_t = dr_{um}/dt \tag{10.4}$$

For this open atmosphere flame, the flame will freely expand to the outmost corrugated surface after the shell of unburned mixture is burned. The flame propagation rate is simply

$$R_t = dr_m/dt \tag{10.5}$$

As the surface area of the flame diverges, care must be exercised when attempting to deduce the turbulent flame speed S_t from this flame propagation rate. In other words, the changing flame surface area prohibits an unambiguous definition of the combustion wave (flame front over time period dt) and consequently, burning velocity. This challenge is particularly true when the flame is small.

We note that turbulence is generally a good thing, as it augments the mass-burning rate from

$$m'_l = r\, S_l\, A \tag{10.6}$$

to

$$m'_t = r\, S_t\, A_{nominal} \tag{10.7}$$

Too much turbulence, however, can lead to partial or complete extinguishment of the flame, as depicted in Fig. 10.4. The slope of normalized turbulent flame speed S_t/S_l versus the normalized turbulence intensity u'/S_l is a function of Lewis number (Le = rate of heat transport/rate of mass transport), Markstein number (Ma = sensitivity of flame speed to

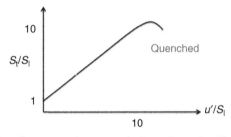

Figure 10.4 Turbulent flame speed versus turbulence intensity. *(Created by D. Ting)*.

stretch), turbulent length scales, turbulence type, etc. Although the exact locations of partial and total quenching are uncertain, some studies (Bedat and Cheng, 1994) showed that S_t continues to increase roughly linearly with u' for S_t (and u') more than ten times S_l. For some cases, as compiled in Lee et al. (2014), this linear trend holds for turbulence intensities beyond 20 times the laminar flame speed!

10.2 RELATIVE SCALES OF FLOW AND COMBUSTION

It is generally accepted that Damköhler (1940) was the first to investigate the relative flow scales with respect to the chemical scales associated with a premixed turbulent flame. The Damköhler number

$$
\begin{aligned}
\text{Da} &= \text{characteristic turbulence time scale/characteristic chemical time scale} \\
&= \tau_{\text{flow}} / \tau_{\text{chem}} \\
&= \lambda / u' / \delta_l / S_l
\end{aligned}
$$

$$(10.8)$$

The Taylor microscale λ is used as the representative length here because it is closely related to the vortical structures in turbulent flows, as portrayed in Fig. 10.5. We see that the larger the turbulence length λ, the longer it takes for the vortex to make a rotation for the same turbulence intensity u'. For a fixed λ, on the other hand, an increase in the intensity u' shortens the rotation duration and hence, the turbulence time scale. The chemical time scale is characterized by the time the reacting front takes to progress (consume) a distance of the laminar flame front thickness δ_l. It is interesting to note that a faster flame has, in addition to a larger value of laminar flame speed S_l, a thinner flame front δ_l.

Let us start with the simpler case where the chemistry is fast relative to the flow. For these large Da turbulent flames, the underlying chemistry, which characterizes the chemical reaction, is largely unaltered by the flow turbulence. As such, turbulence only influences the overall combustion process via the alteration of the otherwise smooth laminar flame surface. In

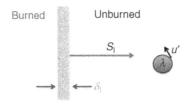

Figure 10.5 Relative flow turbulence and combustion scales. *(Created by D. Ting).*

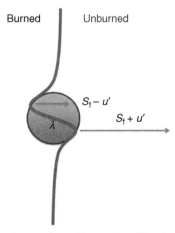

Figure 10.6 A planar flame interacting with a vortex. *(Created by D. Ting).*

other words, flow turbulence tends to wrinkle the flame surface and thus, the global mass-burning rate.

Figure 10.6 illustrates an originally one dimensional, planar laminar flame surface approaching a two-dimensional vortex with characteristic velocity u' at S_f; the flame propagation rate. The vortex twists the flame front such that the portion of the flame surface meeting u' in the negative x direction propagates at a speed of $S_f - u'$. On the other hand, the part of the flame surface encountering u' in the positive x direction has a local flame propagation speed of $S_f + u'$. Therefore, the overall flame propagation rate in the x direction remains unchanged at S_f. However, the initially one-dimensional flame becomes two-dimensional when interacting with the vortex. Most importantly, the flame surface area increases and hence, the total mass-burning rate increases.

We may view these large Da flames as having concave and convex surfaces, though the burning process will consume those surfaces, which are curving into the burned side. The surfaces of the individual positive and negative wrinkles may be assumed to be proportional to the velocity fluctuations $\pm u'$. Therefore, we sense that $S_t \sim u'$, on average.

It is thus clear that the larger the value of u', the more corrugated the flame front is and consequently, the larger the reacting surface area. The reacting surface, however, would tend to consume the corrugation and smooth out the flame front. Nonetheless, the ratio, u'/S_l, determines the degree of flame front corrugation.

Returning back to the flame-vortex interaction depicted in Fig. 10.6, we see that the time required for the flame front to cross the vortex is roughly λ/S_l, where λ is the diameter of the vortex. On the other hand, the time required for the vortex to wrap the flame front around its circumference is roughly $\pi\lambda/u'$. Wrapping the reacting surface more than one revolution around the vortex does not increase the immediate reacting surface area any further. As a result, the time period $\pi\lambda/u'$ is the longest time period required for the vortex to affect the flame to its maximum capability in terms of increasing the reacting surface area. Thence, the asymptote of u'/S_l is approximately π for the ideal case considered. In reality, however, the upper limit can be larger than π, as wrapping beyond one revolution can affect the nearby reacting front.

In 1947, Schelkin (1947) assumed that the entire combustion wave under the influence of flow turbulence is distorted into cones. Thus, Schelkin proposed

$$S_t/S_l = A_{cone}/A_{base} \tag{10.9}$$

where $A_{base} \sim L^2$ and cone height $\sim u' t$, where t $(=L/S_l)$ is the time during which an element of the combustion wave is associated with an eddy of size L moving in the direction normal to the wave. From geometry, the cone area equals cone base times $(1 + 4h^2/L^2)^{1/2}$, where h is the cone height and $h = u'L/S_l$. Therefore

$$S_t = S_l \{1 + (2u'/S_l)^2\}^{1/2} \tag{10.10}$$

We see that this model, crude as it is, predicts the increase in turbulent flame speed somewhat in proportion to the relative turbulence intensity.

10.3 CATEGORIZATION OF PREMIXED TURBULENT FLAME REGIMES

Various classifications have been proposed by different authors. In essence, they are all based on the relative flow-combustion scales. For example, the common categorization is to divide premixed turbulent flame into three regimes:

1. Wrinkled laminar flames

    We have wrinkled laminar flames when the flow, or more specifically, turbulence, is slow compared to the chemistry. Under such condition the smallest flow scale, the Kolmogorov length η, is larger than the laminar flame front thickness δ_l. It is clear that this regime corresponds to the large Da case. The unburned and the burned mixtures separated by a thin flame front are distinctly distinguishable. Yoshida (1988) showed a bimodal probability density function of temperature with well-defined peaks at T_u and T_b.

2. Flamelets in Eddies

"Flamelets in eddies" occurs when the flow turbulence is relatively high such that the flame is moderately affected by the flow; that is, $\Lambda > \delta_l > \eta$. The otherwise continuous laminar flame front is so severely wrinkled by the intense turbulence that it breaks into pieces, which are called flamelets. The underlying chemistry that dictates the local combustion rate, however, is not significantly altered. In terms of Damköhler number, this is the moderate Da regime. This regime has also been referred to as "corrugated flamelet" or "multiple sheet" regime.

3. Distributed Reaction Zone

Distributed reaction transpires when the turbulence is so intense that it directly affects the slower chemistry. With $\Lambda < \delta_l$, the original well-defined flame front or combustion wave no longer exists. Instead, the reaction takes place in a thickened region, which is referred to as the distributed reaction zone. This has also been called "eddy entrainment - combustion in depth flame" (Ballal, 1979).

From the aforementioned discussion, we recognized that the turbulent flame speed may be expressed in terms of Da and Re (Lin, 1996). Based on these two parameters alone, Kido and Huang (1993) formulated a general premixed turbulent flame map, as sketched in Fig. 10.7. With slightly more information, including the relative turbulence-combustion

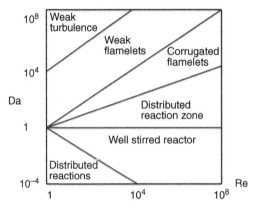

Figure 10.7 Premixed turbulent combustion regimes according to Kido and Huang (1993). *(Created by D. Ting).*

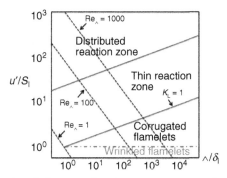

Figure 10.8 Premixed turbulent combustion regimes according to Gülder (1990). *(Created by D. Ting).*

spatial scale ratio, Gülder (1990) generated a somewhat different map as shown in Fig. 10.8. Note that K_L is the Karlovitz stretch factor defined as the local flame residence time over the laminar flame stretching time (Karlovitz et al., 1951).

10.4 TURBULENT LENGTH SCALE AND THE FLAME SURFACE AREA

Let us relax from the single vortex illustration to one with multiple vortex tubes of the same core radius and of unit length. These vortex tubes, which signify the turbulence eddying motions are further assumed to behave as if they are solid rods rotating at a fixed rotation speed. Specifically, the fluid within these vortex tubes undergoes solid body rotation.

Let R be the core radius and Ω the angular velocity. Then the maximum tangential velocity at R is ΩR, and this may be considered to be equivalent to the rms turbulence intensity, u'. The kinetic energy of one of these eddies is

$$KE = \frac{1}{2}\rho \int_0^R (\Omega r)^2 \, d\forall = \rho \Omega^2 R^4 / 4 \qquad (10.11)$$

where KE is the turbulent KE per unit depth or length.

Increasing the core radius from R to $2R$ while keeping the maximum tangential velocity at ΩR and fixing rms turbulence intensity at u', requires a reduction in the angular velocity from Ω to $\Omega/2$. For a unit depth or length

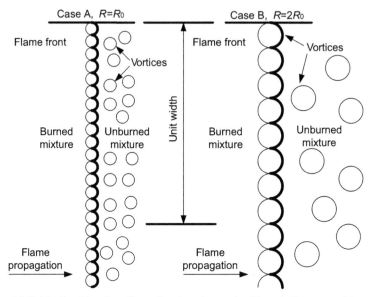

Figure 10.9 Idealized laminar flame fronts saturated with wrinkles caused by small-core and large-core vortices. *(Created by Z. Yang based on Ting [1995]).*

$$\mathrm{KE} = \tfrac{1}{2}\rho \int_0^{2\mathrm{R}} (r\,\Omega/2)^2 \, d\forall = \rho\Omega^2\, R^4 \qquad (10.12)$$

This shows that the turbulent kinetic energy of this larger scale turbulence, where the vortex core radius is $2R$, is *four* times that of the smaller scale turbulence, for the same u'. Therefore, in order to maintain the same turbulent kinetic energy in the two flows, the number of vortex tubes in the smaller-scale turbulent mixture has to be four times the number of vortex tubes in the large-scale turbulent mixture!

10.4.1 A Saturated Wrinkled Flame Front

Figure 10.9 (based on Fig. 2.10 of Ting [1995]) shows an idealized, fully saturated, wrinkled laminar flame fronts, puckered by small-core and large-core vortices. We see that for the fully saturated case, the flame front area per unit width in both small-scale and large-scale turbulent flows is equal to $16\pi R$ per unit depth. In other words, if turbulent eddies are space–filling tubes or the flame surface is saturated as shown in the figure, the change in eddy size does not affect the wrinkled flame front area. Nevertheless, it should be noted that the local curvature and rate of strain effects are larger for the smaller eddy case.

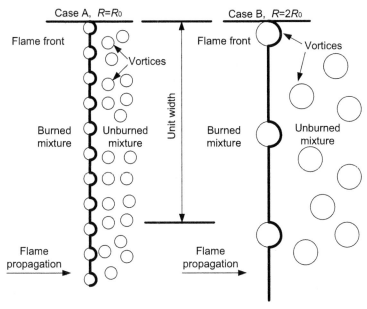

Figure 10.10 Idealized, unsaturated, wrinkled laminar flame fronts; small-core versus large-core vortices. *(Created by Z. Yang based on Ting [1995])*.

10.4.2 An Unsaturated Wrinkled Flame Front

Figure 10.10 (Fig. 2.11 of Ting [1995]) depicts the idealized, unsaturated wrinkled laminar flame fronts in small-scale and large-scale turbulent flows. The total flame front surface area in the unit width shown for the case with eddies of radius R is equal to $(16 + 8\pi)R$ per unit depth. The excess area created with respect to a planar flame front is $(8\pi-16)R$, since the corresponding planar, laminar flame front area is $32R$.

We see that doubling the radius while keeping the turbulent kinetic energy and the turbulent intensity constant results in four times fewer eddies in the flow. Consequently, the total wrinkled laminar flame front area per unit width for the large-scale turbulent flow case is only $(24 + 4\pi)R$ per unit depth. Specifically, doubling the vortex core radius leads to an 11% reduction in the flame surface area. Most importantly, the excess area created by the larger eddies is only $(4\pi-8)R$, which is half of that created by the smaller eddies. In other words, doubling the size of the eddies leads to a 50% reduction in the excess area created.

It should be stressed that the illustration above invokes many idealizations and/or assumptions. Nevertheless, it indicates that smaller scale turbulence is presumably more effective in wrinkling the flame front and creating extra flame surface area than turbulence of larger eddies, for the same turbulence intensity and turbulent kinetic energy.

10.4.3 Comments on Turbulent Length Scale in Combustion

Obviously there are many other factors, which tend to vary with changes in the vortex core size. Here are some points worth noting. First, a turbulent flow always consists of eddies of different sizes; recall the turbulent energy cascade. Second, smaller eddies decay faster than larger ones. Also, smaller eddies lead to higher flame front curvature and higher rate of straining compared to larger eddies. Furthermore, turbulent flows always consist of three-dimensional vortical structures, which can be very different from the ideal two-dimensional case considered.

Premixed combustion studies (Hill, 1988; Hill and Kapil, 1989; Ting et al., 1995) have also indicated that decreasing the size of the eddies can lead to reduced cyclic variations in engines. One probable reason behind this is that the initial flame kernel is convected around by large eddies while wrinkled by the smaller eddies. The bulky convection by the larger eddies is more susceptible to cycle-to-cycle variations in flame growth or burning rate, flame kernel location, and the amount of heat loss to the spark electrodes. In a roughly homogeneously charged engine, these variations in the flame kernel caused by the flow are presumably responsible for the cycle-to-cycle variation of combustion (Hamamoto et al., 1982; Johansson, 1994). It has also been found that engine cyclic variations can be lowered when the early combustion rate is augmented (Mayo, 1975; Stone et al., 1993). In short, besides a more effective enhancement in the burning rate, smaller-scale turbulence also has the tendency to reduce cyclic variations.

10.5 TURBULENT FLAME ACCELERATION AND THE DRIVING MECHANISMS

When igniting a gaseous combustible mixture in a turbulent environment, the turbulent flame speed/turbulence intensity ratio increases as the flame grows. Depending on the chemical and physical parameters involved, this accelerating turbulent flame may develop into a detonation wave. Readers interested in recent research on relatively large scale deflagration (flame propagating at subsonic speed) to detonation (flame propagation at supersonic speed) may start with Groethe et al. (2007), Kim et al. (2013), and Poludnenko (2015). The acceleration tendency of certain turbulent flames also gives rise to the predilection of flashback (flame recessing backward into the feeding passage) in a practical combustion system. To this end, turbulent flame speed seems to be an indicator of the flashback propensity (Lin et al., 2013; Sun et al., 2015).

The following are four possible mechanisms causing the turbulent flame to accelerate. Let us deal with the simple case where the Lewis number is near unity (no preferential heat/chemical species transport), the Markstein number is near zero (the flame speed itself is not sensitive to stretch), and the reaction chemistry is fast compared to the turbulent mixing. This last assumption implies that the mass-burning rate depends primarily on the reacting surface area available.

10.5.1 Progressive Flame-Turbulence Interaction (Evolution Mechanism)

Batchelor (1952) showed that the area of non-reacting surfaces in homogeneous isotropic turbulence increases exponentially with time. Prompted by this, Thomas (1986) argued that the area of a flame also increases with increasing flame-turbulence interaction time, as schematically portrayed in Fig. 10.11. As expected, because a flame continuously consumes its reacting surface, the increase in the reacting surface area of a flame is expected to be slower than that of its nonreacting counterpart (a material surface).

It is important to note that the propagating, reacting flame surface tends to intensify the turbulence just ahead of the flame front (Galyun and Ivanov, 1970; Chew and Britter, 1992). Consequently, the intensified turbulence interacts more strongly with the flame and as a result, the flame propagates faster, and this amplification loop continues until the upper limit, if any, is reached. The flame may become so wrinkled and distorted that pieces of flame are detached from the original flame ball. These pieces of detached burned volume may then grow as flame balls themselves.

10.5.2 Relative Flame/Eddy Size Development

Right after ignition, the length and time scales of the small flame kernels are less than most of those associated with the turbulence (Abdel-Gayed et al., 1984). As such, only eddies which are smaller than the flame can influence the flame front in any significant manner; see Fig. 10.12. The larger eddies convect the flame ball around without affecting the flame front significantly. As the flame grows, the initially larger eddies become

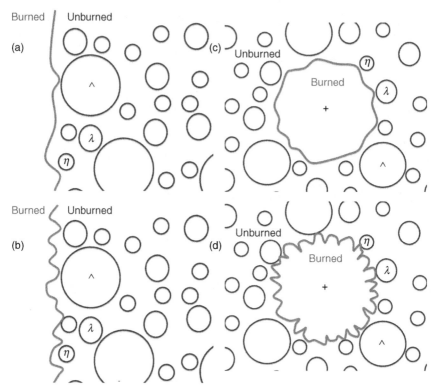

Figure 10.11 The evolution of a reacting surface with time. A planar flame exposed to steady turbulence after (a) a short time, and (b) some time. A flame ball exposed to steady turbulence after (c) a short time, and (d) some time. The cross at the center of the flame ball represents suction, which removes the burned gas at a rate such that the size of the flame remains unchanged with time. *(Created by D. Ting).*

progressively smaller in comparison with the flame ball. As a result, an increasingly larger portion of the turbulence spectrum (energy cascade) becomes effectively involved in corrugating the flame front. When the flame grows larger than the energy-containing eddies Λ, the whole turbulence spectrum is expected to become involved in wrinkling the flame front. Therefore, based on the relative flame/eddy size, it is reasonable to assume that the turbulence would become fully effective in enhancing the flame speed when the flame is an order of magnitude larger than the average eddy size. This saturation asymptote is referred to as the "fully developed turbulent flame." The ratio, S_t/u', which initially increases with increasing flame size, reaches a constant value when the turbulent flame is fully developed.

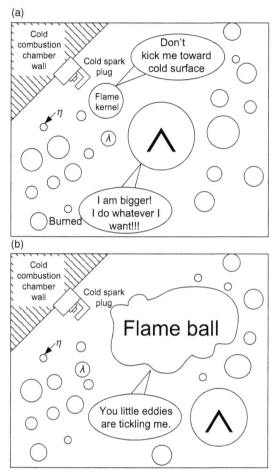

Figure 10.12 Relative flame/eddy size: (a) initially the flame is small compared to turbulence eddies, (b) the flame becomes larger than turbulence eddies. *(Created by Z. Yang).*

10.5.3 Volume Expansion Effect (Expanding-Pushing Mechanism)

As a wrinkled flame grows, volume expansion of a portion of the reacting surface pushes the adjacent reacting surface away from it (Ashurst, 1995). This "expanding-pushing" mechanism can lead to progressively larger degree of flame front corrugation; see Fig. 10.13. The flame surface next to the expanding area experiences the influence of the expansion most directly. The influence diminishes rapidly for the reacting surface farther away from the expanding area. Thus, it appears that the trend of increasing expanding-pushing effect of the reacting surface with the degree of flame

Figure 10.13 The expanding-pushing mechanism. *(Created by D. Ting).*

front wrinkling is only to a certain extent. However, the more wrinkled the flame is, the larger the reacting surface area. The larger reacting surface area bestows more surface for reaction and expansion. The upper limit of this coupled self-enhancing, expanding-pushing mechanism, if it exists, may be reached when adjacent reacting surfaces are so close to each other that they consume each other.

10.5.4 Darrieus-Landau Instability

In 1938, Darrieus (1938) presented for the first time the view that the gas expansion from a concave flame front will result in an increase in the local flame speed, and that the gas expansion from a convex reacting area will lead to a decrease in the local flame speed. This phenomenon, portrayed in Fig. 10.14, gives rise to increasing flame front instabilities and hence, wrinkled flame acceleration. What is depicted in Fig. 10.14 is an overall one-dimensional case, where gas velocities at far left and right are the corresponding unburned gas velocity V_u and burned gas velocity V_b. This hydrodynamic instability was theorized independently by Landau (1944) in 1944. The pertinence of this instability in premixed flame continues to attract copious attention (Steinberg et al., 2009; Creta and Matalon, 2011; Lu and Pantano, 2015).

10.5.5 Attenuation of Flame Front Wrinkling

The only significant attenuating mechanism for a simple flame with no additional effects such as preferential diffusion appears to be the consumption of wrinkles and smoothing of the flame front by the reacting surface itself. It

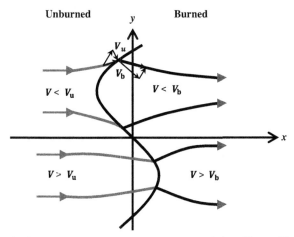

Figure 10.14 The hydrodynamic (Darrieus-Landau) instability. *(Created by P.K. Pradip).*

is still debatable whether this solitary attenuating mechanism will eventually balance out the combined effect of the various augmenting mechanisms. The answer may depend on parameters such as the flame geometry (planar, spherical, or other types of flame) and the type of turbulence (weak, strong, with small or large eddies). It is clear that this asymptote, if any, is beyond the turbulent flame growth period encountered in combustion engines.

10.5.6 Some Progressive Turbulent Flame Growth Evidence

In most engine studies, S_t may level off after some time. This leveling-off is often misinterpreted as the flame reaching the fully developed stage, despite the complications due to changing volume, pressure, and temperature. In fact, the leveling-off of S_t in an engine is mostly due to a decrease in u' (all turbulence decays in the absence of a continuous supply of energy to sustain its ever-prevailing dissipation into heat) and partial flame front quenching by the chamber wall. The following are some examples of progressive flame growth in combustion engines. The turbulent flame speed increases linearly with flame radius with no sign of leveling off for up to 2 cm flame radius in Keck et al. (1987) and up to 4 cm flame radius in Lancaster et al. (1976). Gatowski et al. (1984) found that the surface of the growing flame becomes increasingly distorted with time by the flow turbulence.

Due to the complications of simultaneous variations of numerous parameters involved, engine results are usually difficult to interpret. Idealized flame growth studies, such as constant pressure and/or constant volume combustion, can provide a clearer perception about specific trends by keeping the other parameters fixed. Perhaps the clearest piece of evidence of progressive turbulent flame growth is the experimental study conducted by Palm-Leis and Strehlow (1969). Their spark-ignited, freely-growing, pre-mixed flame ball downstream of a perforated plate appears to be the first concrete experimental evidence illuminating the development of a turbulent flame ball. With integral scales between 1.4 mm to 7.6 mm, the turbulent flames accelerated, with no sign of slowing down, up to 8.4 cm in radius, which was the largest flame size considered. This largest flame size is more than 50 times the integral scale!

Typical values of $(S_t - S_l)/u'$ are plotted as a function of the mean flame radius in Fig. 10.15. These data points are from the tests conducted in a 125 mm cubical combustion chamber where fuel-lean premixed methane-air mixtures were centrally spark-ignited at room temperature and pressure (Ting et al., 1994). Figure 10.15 shows that for the near-unity Lewis number and near-zero Markstein number methane-air flames, $(S_t - S_l)/u'$ increases with increasing r_m with no trend of leveling off up to the maximum mean flame radius $r_m \approx 55$ mm considered.

Haq (2006) extended the experimental developing turbulence evidence and further explained the accelerating turbulent flame in terms of an effective rms turbulence velocity. This effective turbulence velocity is deduced from the integral of the dimensional power spectral density function, from

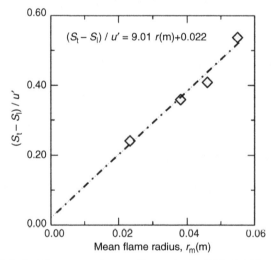

Figure 10.15 Turbulent flame speed versus flame size for 70% stoichiometric methane-air mixture. The premixed charge was centrally spark-ignited in a 125 mm cubical combustion chamber at 296 K and 101 kPa (Ting, 1995).

the highest (Kolmogorov) frequency to the reciprocal of the elapsed time from ignition (Abdel-Gayed et al., 1987). Haq (2006) showed that the effective rms turbulence velocity increases, and so does the turbulent flame speed, with the elapsed time from ignition.

Problems

Problem 10.1. Different regimes of turbulent premixed flame

Based on their experiments, Ballal and Lefebvre (1975) concluded that there are three distinct regions of flame propagation:

Region 1: $u' < 2S_1, \eta > \delta_1$.

In this low-turbulence and low-velocity region, S_t is increased due to wrinkling of the flame. Since all eddies are larger than the laminar flame front thickness δ_1, increasing these eddies increases the wrinkled area and hence S_t.

Region 2: $u' \approx 2S_1, \eta \approx \delta_1$.

This is the region of moderate turbulence in which fresh mixture contains eddies which are both larger and smaller than the laminar flame front thickness. There are two different mechanisms involved in this region:

1. wrinkling of the flame front by eddies larger than flame thickness;
2. increasing the area of interface by eddies entrained in the burning zone.

The first mechanism increases S_t as the scales increase. The second mechanism decreases S_t as the scales are increased. Therefore, these two mechanisms normally cancel out the effect of changing scales.

Region 3: $u' > 2S_1, \eta < \delta_1$.

This is the region of high-intensity and very small eddies. The combustion zone is regarded as a thick matrix of burned gases interspersed with eddies of unburned mixture. The total surface area of eddies is proportional to the inverse of the turbulence scale. Therefore, in this region S_t increases with decreasing eddy size.

Part of the above argument seems to contradict the idea we proposed in this chapter (using two-dimensional, single-size eddies and a planar flame front). Explain.

Problem 10.2. Turbulence and cycle-to-cycle variations

Consider the cycle-to-cycle variations due to fluid motions; see SAE Paper 962084 (Johansson, 1996), for example.

1. Which part (very early period when less than 5% mass is burned, main combustion period where most of the mass is burned, or the burn-out period where the portion along the wall is consumed) of the total combustion period is the most important? Why?

2. How does the size of the eddies affect the cycle-to-cycle variations, i.e., considering turbulences of the same intensity, smaller scale, or larger scale, which leads to lower cyclic variations? Why?

Problem 10.3. Eddy structure turbulent flame growth model

It is shown in Ashurst et al. (1994) that S_t is proportional to $(R/\lambda)(u'/S_l)^{\frac{1}{2}}$. The square root appears to show the slowing down in the increase of S_t with increasing u'. Explain the origin of the square root in the eddy structure model.

REFERENCES

Abdel-Gayed, R.G., Bradley, D., Hamid, M.N., Lawes, M., 1984. Lewis number effects on turbulent burning velocity, Proceedings of the Twentieth Symposium (International) on Combustion, pp. 505–512.

Abdel-Gayed, R.G., Bradley, D., Lawes, M., 1987. Turbulent burning velocities; a general correlation in terms of straining rates. Proc. R. Soc. Lond. Ser. A 414, 389–413.

Ashurst, W.T., 1995. A simple illustration of turbulent flame ball growth. Combust. Sci. Technol. 104, 19–32.

Ashurst, W.T., Checkel, M.D., Ting, D.S-K., 1994. The eddy structure model of turbulent flamelet propagation, the expanding spherical and steady planar cases. Combust. Sci. Technol. 99, 51–74.

Ballal, D.R., 1979. The structure of a premixed turbulent flame. Proc. R. Soc. Lond. Ser. A 367, 353–380.

Ballal, D.R., Lefebvre, A.H., 1975. The structure and propagation of turbulent flames. Proc. R. Soc. Lond. Ser. A 344, 217–234.

Batchelor, G.K., 1952. The effect of homogeneous turbulence on material lines and surfaces. Proc. R. Soc. Lond. Ser. A 213, 349–366.

Bedat, B., Cheng, R.K., 1994. Experimental study of premixed flames in intense isotropic turbulence, Twenty-Fifth Symposium (International) on Combustion. Combust. Flame 100 (3), 485–494, 1995.

Chew, T.C., Britter, R.E., 1992. Effect of flame-induced geometrical straining on turbulence levels in explosions and common burner configurations. Int. J. Eng. Sci. 30 (8), 983–1002.

Creta, F., Matalon, M., 2011. Propagation of wrinkled turbulent flames in the context of hydrodynamic theory. J. Fluid Mech. 680, 225–264.

Damköhler, G., 1940. Der einfluss der turbulenz auf die flammengeschwindigkeit in gasgemischen. Zeitschrift fur Elektrochemie und angewandte Physikalische Chemie 46 (11), 601–626, English translation: NACA Technical Memorandum, No. 1112, 1947.

Darrieus, G., 1938. Propagation d'un front de flame. La Technique Moderne, Paris.

Galyun, I.I., Ivanov, Y.A., 1970. Intensity of turbulence in the combustion zone of a homogeneous fuel-air mixture. Combust. Explo. Shock Waves 6 (2), 211–214.

Gatowski, J.A., Heywood, J.B., Deleplace, C., 1984. Flame photographs in a spark-ignition engine. Combust. Flame 56, 71–81.

Groethe, M., Merilo, E., Colton, J., Chiba, S., Sato, Y., Iwabuchi, H., 2007. Large-scale hydrogen deflagrations and detonations. Int. J. Hydrogen Energ. 32, 2125–2133.

Gülder, O.L., 1990. Turbulent premixed flame propagation models for different combustion regimes, The Proceedings of the Twenty Third Symposium (International) on Combustion, pp. 743–750.

Hamamoto, Y., Wakisaka, T., Ohnishi, M., 1982. Cycle-to-cycle fluctuation of lean mixture combustion in spark-ignition engines. Bulletin of the JSME 25 (199), 61–67.

Haq, M.Z., 2006. Effect of developing turbulence and Markstein number on the propagation of flames in methane-air premixture. J. Eng. Gas Turb. Power 128, 455–462.

Hill, P.G., 1988. Cyclic variations and turbulence structure in spark-ignition engines. Combust. Flame 72, 73–89.

Hill, P.G., Kapil, A., 1989. The relationship between cyclic variations in spark-ignition engines and the small structure of turbulence. Combust. Flame 78, 237–247.

Johansson, B., 1994. The influence of different frequencies in the turbulence on early flame development in a spark-ignition engine, SAE Paper 940990.

Johansson, B., 1996. Cycle to cycle variations in S.I. engines – the effects of fluid flow and gas composition in the vicinity of the spark plug on early combustion, SAE Paper 962084.

Karlovitz, B., Denniston, D.W., Wells, F.E., 1951. Investigation of turbulent flames. J. Chem. Phys. 19 (5), 541–547.

Keck, J.C., Heywood, J.B., Noske, G., 1987. Early flame development and burning rates in spark-ignition engines and their cyclic variability, SAE paper 870164.

Kido, H., Huang, S., 1993. Comparison of premixed turbulent burning velocity models taking account of turbulence and flame spatial scales, SAE Paper 930218.

Kim, W.K., Mogi, T., Dobashi, R., 2013. Flame acceleration in unconfined hydrogen/air deflagrations using infrared photography. J. Loss Prevent. Proc. 26, 1501–1505.

Lancaster, D.R., Krieger, R.B., Sorenson, S.C., Hull, W.L., 1976. Effects of turbulence on spark-ignition engine combustion, SAE paper 760160.

Landau, L.D., 1944. On the theory of slow combustion. Acta Physicochimica 19, 77.

Lee, J., Lee, G.G., Huh, K.Y., 2014. Asymptotic expressions for turbulent burning velocity at the leading edge of a premixed flame brush and their validation by published measurement data. Phys. Fluids 26, 125103.

Lin, C.-Y., 1996. Effects of Da and Re on premixed flame speed. Chem. Eng. Commun. 155, 65–72.

Lin, Y.-C., Daniele, S., Jansohn, P., Boulouchos, K., 2013. Turbulent flame speed as an indicator for flashback propensity of hydrogen-rich fuel gases. J. Eng. Gas Turb. Power 135 (111503), 1–8.

Lu, X., Pantano, C., 2015. Linear stability analysis of a premixed flame with transverse shear. J. Fluid Mech. 765, 150–166.

Mayo, J., 1975. The effect of engine design parameters on combustion rate in spark-ignition engines, SAE Paper 750335.

Palm-Leis, A., Strehlow, R.A., 1969. On the propagation of turbulent flames. Combust. Flame 13, 111–129.

Poludnenko, A.Y., 2015. Pulsating instability and self-acceleration of fast turbulent flames. Phys. Fluids 27 (014106), 1–25.

Schelkin, K.I., 1947. On combustion in a turbulent flow, NACA, TM 1110.

Stone, C.R., Carden, T.R., Podmore, I., 1993. Analysis of the effect of inlet valve disablement on swirl, combustion and emissions in a spark-ignition engine. P. I. Mech. Eng. D-J Aut. 207, 295–305.

Sun, M.-B., Cui, X.-D., Wang, H.-B., Bychkov, V., 2015. Flame flashback in a supersonic combustor fueled by ethylene with cavity flameholder. J. Propul. Power 31 (3), 976–981.

Steinberg, A.M., Driscoll, J.F., Ceccio, S.L., 2009. Temporal evolution of flame stretch due to turbulence and the hydrodynamic instability. Proc. Combust. Inst. 32, 1713–1721.

Thomas, A., 1986. The development of wrinkled turbulent premixed flames. Combust. Flame 65, 291–312.

Ting, D.S.-K., 1995. Modeling Turbulent Flame Growth in a Cubical Chamber, PhD Thesis, University of Alberta.

Ting, D.S-K., Checkel, M.D., Haley, R., Smy, P.R., 1994. Early flame acceleration measurements in a turbulent spark-ignited mixture, SAE paper 940687.

Ting, D.S-K., Checkel, M.D., Johansson, B., 1995. The importance of high-frequency, small-eddy turbulence in spark-ignited, premixed engine combustion, SAE Paper 952409.

Yoshida, A., 1988. Structure of opposed jet premixed flame and transition of turbulent premixed flame structure, Proceedings of the Twenty Second Symposium (International) on Combustion, pp. 1471–1478.

SUBJECT INDEX

Printed in the United States
By Bookmasters